Preparatory Chemistry Modules: A Guided-Inquiry Approach

First Edition

Joe L. March, *University of Alabama at Birmingham*

Craig P. McClure, *University of Alabama at Birmingham*

BROOKS/COLE
CENGAGE Learning

HOUGHTON MIFFLIN COMPANY BOSTON NEW YORK

Introductory Chemistry: A Guided Inquiry Approach
Joe L. March and Craig P. McClure

Publisher: Mary Finch

Acquisitions Editor: Chris Simpson

Assistant Editor: Laura Bowen

Marketing Manager: Nicole Hamm

Marketing Assistant: Julie Stefani

Marketing Communications Manager: Linda Yip

Content Project Manager: PreMediaGlobal

Design Director: Rob Hugel

Art Director: Maria Epes

Print Buyer: Judy Inouye

Rights Acquisitions Specialist: Dean Dauphinais

Cover Designer: Brenton Bartay

Cover Image: Zerega/Getty Images

Compositor: PreMediaGlobal

For product information and technology assistance, contact us at
Cengage Learning Customer & Sales Support, 1-800-354-9706

For permission to use material from this text or product,
submit all requests online at **www. cengage.com/permissions**
Further permissions questions can be e-mailed to
permissionrequest@cengage.com

Student Edition:
ISBN-13: 978-0-8400-6221-5
ISBN-10: 0-8400-6221-4

Brooks/Cole
20 Davis Drive
Belmont, CA 94002-3098
USA

Cengage Learning is a leading provider of customized learning solutions with office locations around the globe, including Singapore, the United Kingdom, Australia, Mexico, Brazil, and Japan. Locate your local office at **www.cengage.com/global.**

Cengage Learning products are represented in Canada by Nelson Education, Ltd.

To learn more about Brooks/Cole, visit **www.cengage.com/brookscole**
Purchase any of our products at your local college store or at our preferred online store **www.cengagebrain.com**

Printed in the United States of America
1 2 3 4 5 6 7 14 13 12 11

Table of Contents

Periodic Table of the Elements

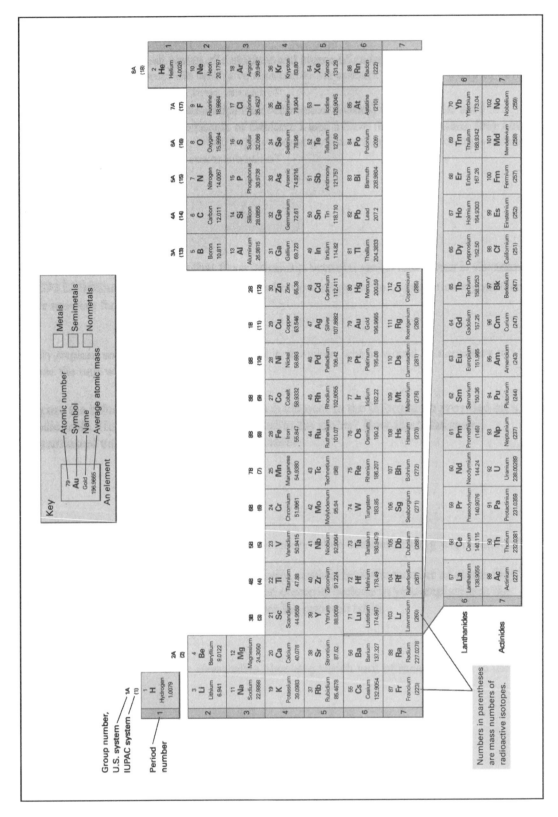

Expanded Table of Contents

Preface

Process-Oriented, Guided Inquiry Learning (POGIL) is a philosophy that there is some process related to how we learn, and we often need discussion with others when we are faced with difficult concepts. The big advances in the POGIL model came about through collaboration between David Hansen and Troy Wolfskill at Stony Brook University; Rick Moog and James Spencer at Franklin & Marshall College; Frank Creegan at Washington College; and Andrei Straumanis at College of Charleston. We have used their work as an important guide as we started this project.

The process part of the work led us to include common sections in all of the activities presented. This process includes the sections:

Objectives

The POGIL model should start with a question; however we adopted the use of the objective as a way to get started. Thus, as a rule, the objectives are a little obtuse. The objective won't give away the conclusion a student is supposed to reach by investigating the model. However, it should give the student a sense of what we expect to find in the "Summarizing Your Thoughts" section. The objectives are placed at the beginning of each activity so that facilitators may quickly assess the content objectives of each activity and how it may fit in with how they wish to present materials in their course.

Getting Started

This section provides some context, but *never* gives away what students are supposed to learn. It might include a reminder of earlier concepts or some historical figures and their contribution to the model, but it won't directly approach the conclusions.

The Model

The model might consist of tables, drawings, brief definitions, or information about an experiment that might not be obvious from a drawing. Students are expected to look at the information and think about what the model might mean.

Exploring the Model

This is the guided section. Do you understand the model presented, and what do you see in the model that might require interpretation? Sometimes these questions are very guided (What is the ratio of x and y?). Sometimes they are quite the interpretation (What do you think happens when…? What can you do with the numbers? Is there something the same in all of the models—why might that be?). This section should develop, and the student should have the sense that there is some logical argument being built or that they are building a set of tools necessary to get to the "Summarizing" section.

Contains questions focused on: how, what, why, compare

May include: draw, write

Summarizing Your Thoughts

This section should *not* introduce new fundamental ideas, but could make connections about the observations of the model in new ways. The student is expected to be able to infer the connection from the model, the exploring sections, or previous chapters/activities. Students are not asked to come up with external concepts.

Students should write something that represents the main ideas that are typically shown as text in other books (definitions, equations, explanations). Ideally these written passages should be good enough to use to prepare for an exam.

Should mostly start with: explain, relate, write, summarize, show, support. (note these are the hard ones to do well, even for us)

Team Skills

As working in groups is an important aspect of this model, a Team Skills section is utilized in some activities to allow student groups to discuss a more challenging problems to further hone their communication skills about a topic, and self-assess the group work. This allows for reflection on the group work, modes for improving each member's contribution in future activities, and identifying difficult concepts for the group which can be a focus of their studies.

End of Chapter Exercises

These exercises are intended for additional opportunities to apply the concepts communicated in each activity. These may be used in the classroom at the facilitator's discretion, or may be assigned as take-home assignments to help reinforce the topics covered. In managing student groups which may work at different speeds, the facilitator could ask faster groups to work on these after completion of the activity in order to keep all groups at a similar pace.

Contains questions focused on: complete, fill in, calculate, describe, write, predict

May include: compare

Acknowledgements

The idea of teamwork was prevalent throughout the preparation of this textbook. The authors thank the many reviewers, facilitators, students, and experienced POGIL practitioners for their contributions and their feedback. The origination of this project began with the help of Jennifer Lewis and K. Kenneth Caswell. All of the contributors have been a tremendous asset during our deliberations about what to cover, how to cover it, and what really happens in the classroom.

This material is based on work supported by the National Science Foundation under Grant No. 0536039. Any opinions, findings and conclusions or recommendations expressed in this material are those of the authors and do not necessarily reflect the views of the National Science Foundation (NSF).

To The Student

Flip through this book. Look at the pages. We hope it looks different from other texts you have used for other science courses. Not too many words, but lots of questions. These questions are designed to help you think about some key issues related to the fundamental concepts used in a chemistry class, and if you do an honest job of working through this book we expect you to learn these concepts at a much deeper level than if we had just told you a bunch of definitions.

Think about it. We can tell you how to ride a bicycle, but will that be enough for you to ride in a race? Just hearing an idea or having someone tell you how isn't sufficient. You need to get in the trenches and ruminate on the idea. Still, just as in learning to ride a bike, you need someone to help support you and guide you as you learn. That is what the questions in this book are about. They should help you figure out a few things and then guide you to write your thoughts down in such a manner that you (and not us, we already kind of have an idea) understand it. Once you write it in your own words it becomes much more valuable to you. Thus, it is imperative that you always try to write in complete sentences, and that you edit your sentences for accuracy because the improper use of a word can lead to big differences in meaning.

This book is also designed to be used by students working in groups. When we work together, we bring in different experiences and have different views. We can talk through issues and bring our unique perspectives to bear on a problem that we encounter. As a group, we are better than we are as individuals since we can bounce ideas around and explain things to each other in new ways. In doing this, it helps us to understand and explain topics from multiple perspectives, which helps us get a better handle on things.

Some of the topics will be familiar to you before you begin, but that same topic may be brand new to your neighbor. It is your responsibility to be respectful in these situations, and to help others learn from your prior experiences. You must find ways to work with others, even if you don't think you have much in common. These interpersonal skills will be more important in your future career than any calculation or problem you will be faced with in this book. So, take advantage of working with your peers. They will have ideas and ways of saying things that will force you to listen critically so that you can learn much more—and it can be fun.

Your instructor's role will also likely be different than you may expect. Throughout this book, we use the term "facilitator" to refer to your course instructor. The facilitator will be around to help, but they won't be doing as much talking as in a normal lecture. They are not present to instruct you on what to do, but help you to discover the big ideas that this book will help you to investigate. They want you to think about the meaning of a picture, a graph, a set of data, or even an equation. You should expect them to ask you even more questions, but those questions should help you recognize how you understand the material. So, don't get frustrated with them.

At the end of this course, we hope that you learn enough chemistry to comfortable and successful in the next course, AND we hope that you also learn that you can look at the material in any text and get more out of it than just a few words. Science isn't about memorization, rather science is about discovery of the world around us. This book is designed to help you make those discoveries, and draw you in to what science is about: teamwork, challenge, and discovery.

Cheers,

Joe March

Craig McClure

To The Instructor

Remember the first time you saw the light bulb go off in a friend's head, and you knew you could teach? Something you said or did allowed another person to catch on quickly. For most of us, that was a very exciting and inspiring moment. The personal satisfaction of helping another person was an influential experience.

As we progress in our careers, we keep expanding our audience. That one-on-one experience became a one-on-many experience. This presents all of us with new challenges. The standard approach to addressing this challenge is to move toward the lecture format. We adapt our ideas into a few sentences or catch phrases that help the larger audience to glean something from what we said, but there is still something missing. We talk in distributions and averages, rather than individual outcomes, in course objectives rather than student goals.

The POGIL approach is designed to provide more of an opportunity to provide one-on-one instruction. By design, the instructor's role changes in a POGIL classroom. You are expected to talk with students; to provide that individual insight; to probe the edges of understanding; and to enjoy helping others. The role shifts from an instructor to a facilitator, from one who tells to one who helps students to discover and describe the answers to their own questions.

The naysayer will note that you aren't at the lectern, so you must not be working. Don't be fooled. Interacting with students and troubleshooting their misunderstandings is much more difficult than reading from your yellowed legal pad (or your PowerPoint file in the modern world). Every class period is different and exciting, and each group of students brings new perspectives. This method of instruction requires an expert facilitator with a deep content knowledge to challenge, inform, question, and encourage each student group and help guide them in each activity.

The POGIL approach may be different, but it still has the same goals we all set out with: students should gain a deep understanding and appreciation for the subject. The POGIL approach views the learning as a process that is suited for deep understanding. Thus, the book is designed to put students through a cycle of work for each topic (see *About the POGIL Model*).

This cycle has been tested in diverse institutions: small private universities, medium sized state universities, and large research-oriented universities. In all of these settings, students have enjoyed the process and learned the material. It is our intention that this book can be used in all of those settings, plus has the potential to be as *flexible* as you want it to be. We have used it as a once-a-week activity, as a short interlude in a large lecture, and as a POGIL-only class. Thus, you can gain as much interaction as you think it needs to be.

As you begin to peruse the chapters and topics, we wish to remind you that we intend for this book to be an introduction to chemistry and not a terminal course. This introduction is designed to provide some very basic concepts and attempts to explore some fundamental ideas that will be explored in greater detail in a future course. We hope that a student that completes the work in an honest manner will be well prepared to succeed in a general chemistry course in a subsequent semester.

Hopefully, you can find the spark of excitement as you use this book that initially attracted you to teaching. I know we did.

Joe March

Craig McClure

1

Jumping Through Science with a P.O.G.I.L. Stick

Chapter Goals:

- Understand the approach taken by the course

- Recognize how your course grade will be determined

- Identify the steps of the scientific method

- Be able to formulate a hypothesis to test experiments in different situations

- Understand how the scientific method is applicable to everyday-life situations

- Be able to identify different approaches to the scientific method

- Establish a set of working rules for the members of your group

ACTIVITY 1.1 EXPLORING THE SYLLABUS

Objectives

- Understand the approach taken by the course
- Recognize how your course grade will be determined

The Model

Refer to your syllabus and this book.

Exploring the Model

1. Why is each member of your group taking this chemistry course, or who is the audience of this class?

2. Will exams be given in this course? If so, how many? When is the first exam?

3. Use the Table of Contents of this book and the syllabus to create a list of topics that are covered in the first five weeks of this course.

4. What activities are graded in this course?

5. What are the required materials (textbooks, calculators, study guides, notebooks, other) for this course?

6. What activity in this course has the greatest effect on your final letter grade?

Summarizing Your Thoughts

7. As a group, discuss the activities associated with this course. Which activity is likely to prove most essential to understanding and achieving your goals in chemistry?

8. Focusing on a grade in a course is a short-term goal. As a group, identify at least three long-term goals for your introduction to chemistry.

9. Review the syllabus and this book and explain how the course is structured for you to be successful in this course and future chemistry classes.

10. An important aspect of this book involves asking you to talk with your classmates. Thus it is important to know your classmates. List each group member's name and one fact about them to help you get to know them better.

ACTIVITY 1.2 THE SCIENTIFIC METHOD

Getting Started

The scientific method is a process for forming and testing solutions to problems or theorizing about how or why things work. It reduces the bias of the researcher and involves experimentation. Once started, the scientific method cycles over and over again to further refine an initial hypothesis.

Objectives

- Identify the steps of the scientific method

- Formulate a hypothesis to test experiments in different situations

- Apply the scientific method is applicable to everyday-life situations

- Identify different approaches to the scientific method

Model

Figure 1.1

Allie's car didn't start this morning. She thought her battery was dead, so she charged it while she was at school. When she came home the battery was charged, so she tried to start the car. It still didn't work. She called her dad to tell him what was going on. He suggested that something else was wrong with her car.

As her initial hypothesis was incorrect, she developed a second hypothesis that her starter wasn't working—*that* was what was wrong with her car.

Exploring the Model

1. Prepare a flow diagram (or a concept map) of Allie's story.

2. Identify the term "hypothesis" in the Model, and work as a group to create a definition for hypothesis in your own words.

3. A key feature of the scientific method is experimentation.

    a. Identify Allie's first experiment.

    b. Explain how the experiment was useful to Allie.

    c. Describe at least one "experiment" that Allie could have performed to confirm that the starter was the problem.

4. Fill in the table using information from the Model and your thoughts about Allie's process.

Table 1.1   The Scientific Method

Steps of the Scientific Method	Actions taken by Allie
Identification of the problem	Allie needed to get to school, but her car would not start.
Hypothesis	
Experimental Test	
Conclusion	A charged battery did not crank the car.
Report Results	

     a. The method in Table 1 is shown in order. Does the method have to be performed in this order? Why or why not?

     b. Describe an everyday decision or event that is similar to an experiment that leads to more than one alternative hypothesis. Explain why some experiments might lead to more than one answer.

Summarizing Your Thoughts

5.   Explain how you use the scientific method in your everyday life without actually realizing you are doing so.

6.   Draw your own conclusions about why the scientific method is important to chemistry.

Team Skills

7.   Discuss with your group the information presented in this activity. Summarize the most important insight gained by any member of the group about the scientific method.

8.   Develop a group answer that describes how the scientific method might be useful when approaching textbook problems in any course.

ACTIVITY 1.3 SETTING RULES OF THE GROUP

Objective

- Establish a set of working rules for the members of your group

Getting Started

Success in your studies is ultimately your responsibility. Instructors have an important role, but in any classroom model the level of knowledge you acquire will depend on your ability to personally work through the key concepts of the material.

Still, you won't have to go through this course alone. We will be utilizing the talents and abilities of your classmates throughout the semester. However, for this to be successful everyone has to take some responsibility. There are many ways to help people work together, and your instructor may require you to establish formal roles (leader, recorder, reflector, or similar roles). However, this activity is designed to help you think about some very basic responsibilities related to working with your classmates.

The Model

Contract for Effective Group Work

- All members of the group will contribute to each and every activity.

- If a member of the group does not understand the material, the members of the group that do understand will work to explain the material to their classmate.

- Each group member shall be free to express their opinion without fear of ridicule from any of the other group members.

- Each group member will do everything in their power to attend every class so that they may help their classmates and themselves learn the material.

- Each member of the group is responsible for making sure that the other members of the group are working and staying on task. They are also responsible for making sure that no one person dominates the discussion.

Signed

The Members of the Group

Exploring the Model

1. How many members of the group are responsible for contributing to all of the activities?

2. Who is responsible for signing the Contract for Effective Group Work?

3. Think of a time when working in a group is effective (sports, emergency responders, military forces). Do these groups have an implied group agreement?

4. Select an example of an effective group, and write a "Contract for Effective Group Work" for that group.

5. How do groups hold members responsible for behaviors identified in a contract?

6. Prepare a list of behaviors that ideal students exhibit when working in a group.

7. The contract states that all members of the group are *responsible for making sure the workload is distributed evenly*. How can you make sure this is happening?

8. What actions should be taken if a group member decides to miss a class?

9. Many times a member of a group will not ask a question about something they do not understand to avoid being made fun of or feeling less intelligent than the other members of the group. What steps can you take as a group to make everyone feel comfortable enough to ask a question? Are there other ways to assure that everyone understands the topics that are being discussed?

Summarizing Your Thoughts

10. As a group, write your own contract that can be used to successfully navigate this course. Be sure to list in your contract what actions should be taken if members of your group fail to live up to the contract.

End of Chapter Exercises

1. A simple syllabus might require three 100-point exams and one 150-point cumalative final exam. What percentage of the total points is the final exam?

2. In the chemistry course at Big State U, it is possible to earn 650 points through exams, homework, and class participation. If each homework set contains 10 problems worth 1 point each, what percentage of the overall grade is

   a. A homework set?

   b. A single problem on the homework set?

   c. Discuss with your group when it is appropriate to ask the instructor to re-grade a single homework problem, and summarize your discussion below.

   d. A student misses a homework set and asks to submit the homework after the due date. Under what conditions should an instructor accept the late work?

3. Create a story about a situation that is familiar to you that requires you to collect some information and make a decision. Your story should contain all of the aspects of the scientific method you identified.

4. Use Table 1.2 to organize your story and to clarify how you have related it to the scientific method.

Table 1.2   Organizing Your Story

Steps of the Scientific Method	Actions in Your Story

5. Identify which part of the scientific method each statement is associated. Provide a brief explanation of your thoughts if a statement can be classified within more than one step of the scientific method.

a. There is a direct relationship between mass and volume.

b. There are 14 boys and 21 girls in class today.

c. How does a calculator work?

d. Energy drinks increase the risk of high blood pressure.

e. In a study of 5,000 undergraduate students, we found that most students enjoyed their chemistry class.

f. Students that do more homework make better grades.

2

Representing the Little Things in Life

Getting Started:

One of the challenges in chemistry is that we must possess the ability to think about chemistry from four different perspectives: a macroscopic view, a descriptive vocabulary, a symbolic representation, and a nanoscale view. This ability is similar to learning four different languages. You have to be fluent in all four and able to think about how the other three perspectives might be useful for every chemical process. This chapter will introduce the four perspectives and explore how chemists use these different ways of thinking.

Chapter Goals:

- Recognize the importance of symbols or abbreviations

- Identify an approach to understanding unfamiliar abbreviations

- Recognize the symbols of the periodic table

- Use chemical symbols from the periodic table to represent molecular formulas

- Recognize a nanoscale picture and be able to represent it with a symbol or descriptive vocabulary

ACTIVITY 2.1 USING SYMBOLS

Objective

- Recognize the importance of symbols or abbreviations

- Identify an approach to understanding unfamiliar abbreviations

Getting Started

Every day we use symbols, words, and drawings in various ways to show the same thing. Our initial model begins with common abbreviations and symbols so that we can begin to think about the importance of these symbols or abbreviations.

The Model

Table 2.1

Macroscopic View	Descriptive Vocabulary	Property Measured by Unit	Symbolic Representation
	Dollar	Monetary value or cost	$
	Penny	Monetary value or cost	¢

Exploring the Model

1. What is the symbolic representation for Dollar? _____

2. What is the descriptive vocabulary word that is represented by the symbol ¢? _____

3. Both the dollar and the penny represent money. How are these two related? Describe their relationship in words, and then with a mathematical relationship.

4. Are all symbols simply an abbreviation of the vocabulary word or phrase? _____

5. Provide an example of a symbol that is not simply an abbreviation of the vocabulary word.

6. Prepare a list of at least three other everyday quantities that use symbols to represent a unit of measure.

7. Why would you choose to use symbols instead of the entire word?

Summarizing Your Thoughts

8. Write a short paragraph that explains the relationships among the macroscopic view, the descriptive vocabulary, and the symbolic representation.

9. Describe how visualizing the macroscopic view could be helpful when calculating any new quantity.

10. What is a resource that you could consult when you see a new symbol that you do not understand?

Team Skills

11. Was your group useful in helping everyone recognize the symbols presented? What could be done to improve the quality of the group work as you move into the next section?

12. Consider the following sequence: 5, 5°, 5°C. Discuss and come to a group consensus answer as to why the last value in the sequence is more descriptive than the previous ones.

ACTIVITY 2.2 THE PERIODIC TABLE

Objective

- Recognize the symbols of the periodic table

Getting Started

The periodic table is an extremely valuable tool for the chemist. We will explore the organization of the table and the properties of the symbolically represented elements in more detail later. However, for now we want to recognize that the periodic table is an arrangement of the symbols that we use in chemistry to represent the elements. We use the names of the elements many times in our daily lives. We have heard of oxygen for the air we breathe and potassium as an important part of our diet. You will learn what it means to be an element and the properties of elements in detail later, but initially we will need to learn how to quickly interchange an element's name and symbol so that you can have a reasonable understanding of more complex ideas. Explore the Model to see the common traits among the symbols in this table.

The Model

Figure 2.1   The Periodic Table, Main Group and Transition Elements

Exploring the Model

1. How many elements' symbols contain just one letter? _____

2. How are the elements arranged, with respect to the number printed above each chemical symbol?

3. What case (upper or lower) is the first letter of every chemical symbol? _____

4. For the elements with more than one letter, what case (upper or lower) is the second letter of every chemical symbol? _____

5. From the Table, what is the most common number of letters used to represent an element?

   _____

6. Some elements' symbols are derived from Latin or from the discoverer's native language. Identify the three elements in Figure 2.1 whose chemical symbol does not appear to be related to the name of the element.

7. Write the chemical symbol for all the elements that have a symbol that begins with the letter A. Is there a simple way to associate the name with the second letter in each of these symbols?

Summarizing Your Thoughts

8. Prepare a statement that helps you remember the rules for capitalization of the letters in an element's symbol.

9. What would be some likely places to find a periodic table if you needed one to look up symbols of the elements?

Team Skills

10. What insight have you gained as a result of your team's performance today?

11. Explain why writing a chemical symbol correctly is important when communicating about chemistry within your group.

ACTIVITY 2.3 MOLECULAR REPRESENTATIONS

Objectives

- Use chemical symbols from the periodic table to represent molecular formulas

- Recognize a nanoscale picture and be able to represent it with a symbol or descriptive vocabulary

Getting Started

The next level representation used in chemistry is a chemical formula. A chemical formula represents an individual molecule or compound. A compound is any group of atoms that are bound together in a regular pattern. A molecule is the smallest particle that can be obtained with the same characteristics as the bulk compound. If we break up a molecule, we can end up with atoms, but atoms have different characteristics than the molecule. There are an infinite number of chemical formulas made up from over 100 known elements. There are different types of chemical formulas that we will introduce later. First we will explore molecular formulas.

Molecules and atoms are much too small for us to be able to see, and if we were able to see them, they would most certainly not look like any of these figures. The different representations of a molecule help us understand how molecules behave and provide us with a way to understand and visualize their behaviors. Through the years, our ideas of what molecules may look like have advanced, but it is still impossible to accurately represent what molecules look like. So we rely on chemical symbols or models that we *can* examine to try to understand why molecules behave as they do.

We start with the water molecule. There are many different representations of a molecule of water, and each shows a different characteristic of water. We will probe these representations to see the value of the different representations and then practice interchanging the nanoscale representation with the symbolic.

Our day-to-day interactions with matter occur on the macroscopic level, which describe observations made with the naked eye. Macroscopic observations about water would be that water is a clear and colorless liquid. The nanoscale or particulate-level describes water as we might "see" it; though it is not possible to directly observe water molecules with our eyes. The pictures shown are drawn on the basis of experimental data, and represent that a water molecule is composed of one oxygen atom and two hydrogen atoms, and the drawings are starting to tell us something about the shape of the molecule. This introductory activity will help you determine how to relate the drawings to chemical formulas.

The Model

Figure 2.2   Different representations of a Water Molecule

Table 2.2 Relating The Nanoscale to Molecular Formulas

Ball-and-Stick Model	Molecular Formula	Atoms of First Element	Atoms of Second Element
	CH_4	1	4
	H_2O	2	1
	NH_3	1	3
	HCl	1	1

Exploring the Model

1. What is the shape used most often to represent the individual atoms in this model? _____

2. How do we show that elements are connected in the Model?

3. Do all of the elements appear to be made of atoms of the same size? _____

4. Is the hydrogen atom in Table 2.2 represented by a white or a grey sphere?

5. How many hydrogen atoms in

    a.   HCl _____   b. NH_3 _____   c. CH_4 _____   d. N_2H_4 _____

6. Look at the three representations of water presented in Figure 2.2. Sketch the representation that is the most useful for describing:
    a.   the volume of the molecule.

    b.   the H-O-H angle.

    c.   the connections (bonds) between each atom.

7. How are the number of atoms of each element communicated in the molecular formula?

8. Is a subscript used when there is only one atom of an element in the compound?

Summarizing Your Thoughts

9. What can we say about what a chemical formula tells us about the ratio of one element to another? Use an example from Table 2.2 to justify your answer.

Team Skills

10. As a group, discuss and compose a statement describing the difference between an element and a compound.

End of Chapter Exercises

1.   Where or when do symbols cause confusion?

2.   Ounces can be converted to pounds, and feet can be converted to inches. Why can't ounces be converted to inches?

3.   Given that 1 zloty = 1.13 litas = $ 0.35   What property is the zloty used to measure?

4.   Given 40 falls = ¼ Scottish acre = 0.375 roods   What property is the rood used to measure?

5.   Explain why writing the symbol SN for tin (Sn) would cause confusion.

6.   Table 2.3 contains the 20 most common elements found on earth. You will encounter these elements throughout your studies and in your daily life; thus it is useful to learn to relate the names and symbols for these elements. Use a periodic table to complete the table:

Table 2.3 The Twenty Most Common Elements on Earth

Element	Symbol
Hydrogen	
	He
Neon	
Fluorine	
Oxygen	
	S
Carbon	
	N
Aluminum	
Sodium	
Potassium	
Iron	
	Ni
Copper	
Iodine	
	Cl
	Pb
Silicon	
Bromine	
Magnesium	

7. Give the number of atoms of each element present in one molecule of each of the following compounds.

Table 2.4 Interpreting the Subscript

Compound Name	Molecular Formula	Atoms of First Element	Atoms of Second Element
Dihydrogen sulfide	H_2S		
Dinitrogen tetroxide	N_2O_4		
Carbon dioxide	CO_2		
Carbon monoxide	CO		
Carbon tetrafluoride	CF_4		

8. Explain the difference between the representations 2 Ar versus Cl_2. How does the position of the numeral 2 affect one's thinking?

9. Complete Table 2.5 by writing the molecular formula for each compound.

Table 2.5   Writing Molecular Formulas

Compound Name	Molecular Formula	Atoms of First Element	Atoms of Second Element
Nitrogen monoxide		1	1
Sulfur dichloride		1	2
Sulfur trioxide		1	3
Boron tribromide		1	3
Dicarbon hexahydride		2	6
Diphosphorus pentoxide		2	5

10. Draw a nanoscale representation of sulfur dioxide and sulfur trioxide. Make sure your drawings are clearly labeled.

   a. In your drawings, how did you decide what connections to make?

   b. What other information would have been useful to know when you started to draw SO_2 and SO_3?

3

So You Want to Mix Things Up? (States of Matter and Mixtures)

Chapter Goals:

- Identify the three states of matter (solid, liquid, gas) from a nanoscale drawing

- Describe how the distance between molecules affects simple properties

- Understand how the placement on the periodic table can help identify whether an element is a metal or nonmetal

- Show a symbolic representation of a mixture on the nanoscale

- Make a determination between two types of mixtures

ACTIVITY 3.1 NANOSCALE VIEW OF THE STATES OF MATTER

Objective

- Identify the three states of matter (solid, liquid, gas) from a nanoscale drawing
- Describe how the distance between molecules affects simple properties

Getting Started

We are very familiar with the three states of matter. In our everyday lives we come into contact with all three forms: solid, liquid, and gas. In this activity, we will look at a simplified view of molecules in the three states of matter. Our simplification involves showing the molecules in only two dimensions. Thus, your observations about how close the molecules are to each other needs to be extended beyond the dimensions of this page. Still, it is expected that you can take these simple pictures and create a way of thinking about how the distance between the molecules will affect simple properties.

The Model

Table 3.1 Three Views of Water

	Gas	Liquid	Solid
Looking at elemental bromine, Br$_2$			
Density	0.0064 g/mL at 300K	3.12 g/mL at 293 K	4.05 g/mL at 123K
Shape	Will take the shape of the container	Will take the shape of the container	Will form its own shape and be rigid or fixed
Volume	Takes the volume of the container	Has a constant volume independent of the size of the container	Has a constant volume independent of the size of the container
Particle Motion	Random motion throughout the container	Random motion throughout the container	Local vibration in a fixed position, no long-range motion

Exploring the Model

1. In which state of matter are the bromine molecules farthest apart?

2. What is meant by *local vibration* in the description of the particle motion of a solid?

3. Consider how the three states of matter relate to their container. Describe what you would expect to see in the macroscopic view if you were to try to fill a balloon with each state of matter.

4. Do your macroscopic expectations of filling a balloon agree with the Model's description of shape?

5. Imagine you fill a piston (or a syringe) with a gas. Would you expect to be able to move the piston?

6. Would you be able to move the piston if it were filled with rocks?

Summarizing Your Thoughts

7. In the drawings included in the model, what distinguishes a gas from a liquid?

8. Explain how the relative positions (how close they are to each other) of the molecules or atoms at the nanoscale affect your ability to move the piston in Questions #5 and #6.

9. Players on a soccer team could be found in three places: 1) on the field playing, 2) on the bench sitting with their team, or 3) back at their respective homes. Explain which state of matter could be described by each situation.

10. Discuss this analogy within your group. Are there problems with the analogy of the soccer team to the states of matter? Explain where this analogy might not be appropriate.

ACTIVITY 3.2 METALS AND NONMETALS ON THE PERIODIC TABLE

Objective

• Understand how placement on the periodic table can help identify whether an element is a metal or a nonmetal

Getting Started

The periodic table can be a very valuable tool for the chemist. In chapter two we saw how each block represents a different element. We will continue to explore how we can continue to gain more insight from the periodic table as we proceed, but for now let us take a look at how those elemental symbols are arranged on the periodic table to help identify different characteristics of the element.

The Model

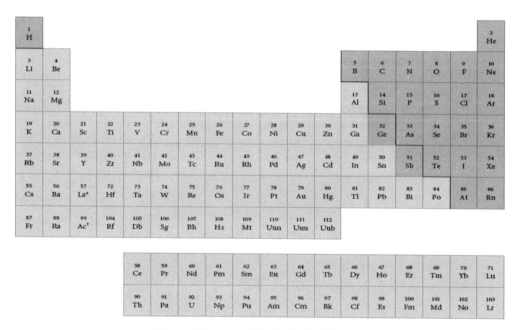

Figure 3.1   The Periodic Table

Exploring the Model

1. Circle the elements Fe, O, and Al on the periodic table above.

2. Describe the macroscopic properties (the properties you can see with your eyes) of iron (Fe).

3. Describe the macroscopic properties of elemental oxygen (O_2) as found in the air.

4. Review the elements around iron on the periodic table. Are their macroscopic properties similar to iron?

5. Review the elements around oxygen on the table. What do you know about their macroscopic properties?

6. What shading is being used to describe nonmetals?

7. What shading is being used to describe metals?

Summarizing Your Thoughts

8. The shading chosen in for Figure 3.1 is not shown on all versions of the periodic table. Write yourself a note that describes the side of the periodic table where metals are likely to be found.

9. Where would you expect to find elements that are nonmetals?

ACTIVITY 3.3 REPRESENTING MIXTURES

Objectives

• Show a symbolic representation of a mixture on the nanoscale

• Make a determination between two types of mixtures

Getting Started

Now that we understand the different types of matter, and we are familiar with how the periodic table can help us to understand what type of material we have, we can begin to look at what will occur when we combine compounds that are in different states.

The Model

A student is given two beakers of water. In the first beaker he adds a large scoop of sugar, and in the second he adds a large scoop of sand. He stirs them up for a minute and then reports what he sees.

Sugar beaker: He could no longer see the sugar, and the beaker of water looked clear.

Sand beaker: The sand settled in the bottom of the beaker, and the water above the sand looked clear.

Bowl of sugar   Bucket of Sand

Figure 3.2   Sugar and Sand

Table 3.1

Macroscale	Descriptive Vocabulary	Formula of the Compounds	Nanoscale	Key
	Sugar and Water	$C_6H_{12}O_6$ & H_2O		water sugar
	Sand and Water	SiO_2 & H_2O		SiO_2 (simplified representation)

Exploring the Model

1. Are the sugar molecules distributed within the water?

2. Is the SiO_2 distributed in the same manner?

3. Describe the macroscopic differences between the two mixtures.

4. Describe how the nanoscale pictures differ.

5. Scientists classify the two models shown using the nomenclature *heterogeneous* and *homogeneous* mixture. The two solutions described in the model represent these two types of mixtures. Consider the prefix to these two words, discuss with your group, and explain how to distinguish between a homogeneous and heterogeneous mixture.

Summarizing Your Thoughts

6. In the space below, give a nanoscale description of a mixture in which a solid material is present in a glass of water.

7. In no more than three sentences summarize what was learned about mixtures today.

End of Chapter Exercises

Activity 3.1

1. Water is an exception to the model presented in Activity 3.1. The density of solid water is less than the density of liquid water. What macroscopic observation confirms that water is an exception?

2. Liquid nitrogen is used to make fancy drinks in some restaurants. Write a few sentences and draw pictures that would explain the change of state of nitrogen, as it goes from a liquid to a gas.

Activity 3.2

3. Look at the periodic table to see how chlorine might be related to bromine. Chlorine is a gas at room temperature, and its formula in this state is Cl_2. Sketch a drawing to represent chlorine gas.

4. You have not likely ever seen elemental sodium, but based on the position of sodium on the periodic table would you expect it to appear more like iron or like oxygen? Explain your choice.

Activity 3.3

5. After observing the beakers the student was instructed to take a cup and remove some water
    from the top portion of the sugar beaker. He then took a cup and removed some water from
    the top portion of the sand beaker. Sketch a nanoscale view of the liquid that was removed
    from each beaker.

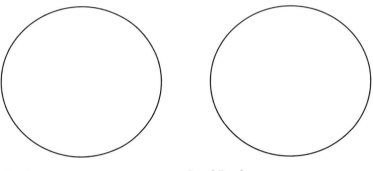

        Sugar Beaker   Sand Beaker

6. Based on what you have learned, predict whether the following mixtures would be
    <u>heterogeneous</u> or <u>homogeneous</u>:

    a. Orange juice (with no pulp)   _____

    b. Clear ocean water   _____

    c. A piece of raisin bread   _____

    d. The air inside a helium balloon _____

    e. A pond with some algae   _____

4

Numbers and Units

Chapter Goals:

- Describe why units are important in communicating information
- Identify common units
- Make rational choices about which units to use
- Identify equalities and use them in conversion factors
- Be able to describe the relationship between the magnitude of a measurement and the unit used to describe that measurement
- Know the SI units of measurement
- Know what the prefixes represent in the SI system
- Be able to use powers of ten to describe the magnitude of a number
- Know what differentiates base units from derived units
- Given a derived unit, know how to combine measurements to obtain it
- Understand what density represents, and be able to carry out calculations using density

ACTIVITY 4.1 THE IMPORTANCE OF UNITS

Objectives

- Describe why units are important in communicating information

- Identify common units

- Make rational choices about which units to use

Getting Started

As chemists we deal with numbers everyday. For example, numbers are a useful way for a forensic chemist to describe a toxic dose of arsenic, or for an industrial chemist to describe how much citrus peel will be needed to meet customer demand for sweet-smelling dog shampoo. Since chemists so often communicate with numbers, they have developed standards for communication that help to prevent misunderstandings. Fortunately, as you'll discover in this activity, these standards are "common sense" and you probably use the same ideas in your daily life.

The Model

You are driving to find a friend's house, and get lost. You stop and ask a stranger walking his dog for directions, and he seems to know where you are going. He says:

"Take a left at the next stoplight, and then go five."

Exploring the Model

1. What direction did the man tell you to turn? _____

2. How far did the man tell you to drive? _____

3. What is it about these directions that might cause confusion?

4. What question would you ask of the man in order to clarify these directions?

Summarizing Your Thoughts

5. Complete the table using units of measurement that the members of your group might use every day. The first row is completed for you.

Table 4.1

Types of Measurement:	Common Units of Measure:		
Length of a pencil	Inches	or	Centimeters
Length of a soccer field		or	
Time spent studying chemistry		or	
Time spent in high school		or	

6. Is *pounds* a more appropriate unit to communicate the mass of a pumpkin or the length of a French fry?

7. Is *teaspoons* a more appropriate unit to communicate the amount of sugar in a cup of coffee or the amount of corn grown in Iowa annually?

8. Based on your responses above, what factors should you consider when choosing which unit of measurement to use in a particular situation?

Team Skills

Discuss as a team, and use complete sentences and proper grammar to communicate your team's consensus answers for the questions below.

9. Briefly explain using the examples you listed in the table, why it would be necessary to include the unit when referring to time.

10. Why is including the unit of measurement necessary when communicating with numbers?

11. What do chemists need to think about when choosing which unit of measurement to use?

ACTIVITY 4.2 CONVERSION FACTORS AND MAGNITUDE

Objectives

- Identify equalities and use them in conversion factors

- Be able to describe the relationship between the magnitude of a measurement and the unit used to describe that measurement

Getting Started

From the previous activity we have seen that it is necessary to include the unit of measurement when trying to communicate with someone about a measurement. But what do you do if people are unfamiliar with the unit of measurement you want to use? We will now see how some units of measurement can be related to each other through the use of an *equality*, or *conversion factor*.

An equality is a mathematical statement of the relationship between two different units of measurement. A conversion factor is a rearranged form of the equality that can be used to convert one unit of measurement in the equality to the other by simple multiplication. We start with an unfamiliar equality so that you can focus on how you process this information in order to create a conversion unit.

The Model

A pastry chef needs 12 *firkins* of Key lime juice to make pies for a banquet, but Key lime juice is only sold in *pins*.

$$\text{The equality: } 2 \, pins = 1 \, firkin$$

The conversion factor for converting pins to firkins: $\dfrac{1 \, firkin}{2 \, pins}$

The conversion factor for converting firkins to pins: $\dfrac{2 \, pins}{1 \, firkin}$

To convert 42 pins to firkins, we can use: $42 \, pins \ \times \ \dfrac{1 \, firkin}{2 \, pins} \ = 21 \, firkins$

Exploring the Model

1. How many pins are in a firkin? _____

2. How many pins are in 12 firkins? _____

3. Write the conversion factor that you used to determine how many pins are in 12 firkins.

4. How are the conversion factors for converting pins to firkins and for converting firkins to pins related?

5. Is $\dfrac{1\,firkin}{2\,pins}$ equal to $\dfrac{2\,pins}{1\,firkin}$? Discuss as a group and explain your answer.

6. Is $\dfrac{1\,firkin}{2\,pins}$ is equal to 1? Discuss as a group and explain your answer.

7. Examine the example in the model of converting 42 pins to firkins. Mathematically, what happens to the units "pins" if it is in the first value, and is on the bottom of the conversion factor? (Why does the unit pins not occur in the answer?)

8. Which unit represents a larger volume, a pin or a firkin? If you convert from pins to firkins, would you expect the numerical portion of the measurement to increase or decrease with the conversion?

Summarizing Your Thoughts

9. Use your answers above to explain how conversion factors can be used to convert one unit of measure to another with simple multiplication.

Some Common Equalities:

12 inches is the same as 30.5 cm
1 pound is equal to 16 ounces
0.0005 ton = 1 pounds
3.1 miles is the same as 5 kilometers
There are 60 seconds in one minute

10. From the table above of common equalities, list the key words or symbols that are used when identifying an equality.

11. Given that 12 inches = 1 foot = 30.5 cm. If a distance is measured using all three of these units, which would give the largest numerical portion of the measurement?

12. In converting from one system of measurement to another, the number changes and the unit changes. Does the measurement itself change? Why or why not?

13. Describe how any equality can be turned into a conversion factor.

14. Explain how the equalities and the resulting conversion factors can help you estimate answers before you begin using a calculator.

15. Discuss your answers above with your classroom facilitator, and have them initial below.

Facilitator Initials: _____

Team Skills

16. The tool that you have been using to solve problems properly (in paying attention to units) is called *dimensional analysis*. As a group, establish a working definition for dimensional analysis.

17. Briefly describe how well your group is working together. Be sure to include a specific example of a contribution that each member of your group has made. (Understand that your instructor will compare the answers you give with what they observed in class.)

ACTIVITY 4.3 SI UNITS AND PREFIXES

Objectives

- Know the SI units of measurement

- Know what the prefixes represent in the SI system

- Be able to use powers of ten to describe the magnitude of a number

Getting Started

In scientific inquiry, units from the SI system (International System) or metric system are primarily used for measurement. This standardized set of units provides everyone with a common, fundamental measurement system. In this way, we can all communicate easily with each other.

Throughout science, we work with very big and very small measured values. A value is cumbersome when we use all of the decimal places and do not know how accurate it is. Therefore, we use a system of prefixes to communicate a common multiplication factor.

In this activity, we will introduce the common SI units and the multiplication factors that you will likely encounter in other science courses. In addition, we hope that through the use of these SI units you will gain a greater appreciation for the magnitude of the nano-scale.

The Model

Table 4.2 SI Units and Some Equalities

Measured Quantity	SI Unit (and symbol)	Equalities with Other Units
Length	meter (m)	1 meter = 3.2808 feet
Volume	Liter (L)	1 m^3 = 264.2 gallons 1 Liter = 0.2642 gallons
Time	second (s)	60 s = 1 minute
Mass	kilogram (kg)	1 kg = 2.2046 pounds
Temperature	Kelvin (K)	K = °C +273.15 °C = (°F - 32) * (5/9)

Table 4.3 contains the most common prefixes used in the SI system. There are others listed in various sources, it will be beneficial to memorize each of these and their magnitude.

Table 4.3

Prefix	Multiplication Factor		Symbol	Verbal Description
kilo	1000	10^3	k	One thousand units
deca	10	10^1	dk	Ten units
	1	10^0		One unit
deci	0.1	10^{-1}	d	One tenth of a unit
centi	0.01	10^{-2}	c	One hundredth of a unit
milli	0.001	10^{-3}	m	One thousandth of a unit
micro	0.000001	10^{-6}	μ	One-millionth of a unit
nano	0.000000001	10^{-9}	n	One billionth of a unit

Exploring the Model

1. Which prefix in the table corresponds to the smallest division of a unit? _____

2. What is one of the multiplication factors for milli-? _____

3. What are the two different representations used to describe Multiplication Factor in Table 4.3?

4. Looking at Table 4.3, is centi- less than or more than one unit?

5. Would kilo- most likely be used to represent a large amount of something or a very small amount of something?

Summarizing Your Thoughts

6. Write down four different units for the SI unit of length by adding prefixes to the base unit.

7. How is the standard notation of the Multiplication Factor (second column) related to its notation in powers of ten (third column)?

8. Write the steps necessary to show how many kilograms are in a milligram using only words (no numbers).

Team Skills

9. Discuss and decide as a group how it is beneficial to use prefixes based on powers of ten to describe very large and very small numbers, rather than the unit without the prefix. Use complete sentences and proper grammar to record your consensus response.

ACTIVITY 4.4 DERIVED UNITS AND DENSITY

Objectives

- Know what differentiates base units from derived units

- Given a derived unit, know how to combine measurements to obtain it

- Understand what density represents, and be able to carry out calculations using density

The Model

Diamond and graphite are both made up of carbon atoms, but differ on the nanoscale (what we imagine these materials to look like on the atomic level). A diamond with a mass of 5 grams has a volume of 1.43 mL. A piece of graphite with a mass of 5 grams has a volume of 2.27 mL.

Exploring the Model

1. In the numerical data presented, what is the *same* about the diamond and graphite used in this example?

2. In the numerical data presented, what is *different* about the diamond and graphite used in this example?

3. In working with measurements, some values cannot be directly measured, but must be given values with a combination of different (or the same) units. Density is one of those units, which is defined as mass per unit of volume. From the Model, identify which units describe mass and volume for the diamond.

    mass –

    volume -

4. In stating "mass per unit of volume", what does the *per* mean?

5. Some units may be measured directly (such as feet, grams, seconds, etc.) by comparison to a standard measure (such as a tape measure, balance, clock, etc.). These are called *base units*. Other units are not directly measurable, but are quantifiable through combinations of other units. These are called derived units. In the table below, state which base units may be combined to measure the given value.

What you want to measure	Base Units Combined to Measure it
Price of a gasoline	
Speed of your car	
Price of meat in the grocery store	
Density	
Liter (it is a derived unit! See Table 4.2 for a clue)	

6. Calculate the density of a diamond, and the density of graphite in $^g/_{mL}$ from the information given in the model. Show your work.

7. Data on a series of woods is shown in the table below. Complete the table for each type of wood.

Material	Mass	Volume	Density
Red Oak	6.95 g	10.59 mL	
Balsa		6.80 mL	0.17 $^g/_{mL}$
Ironwood	5.15 g		1.23 $^g/_{mL}$
Eucalyptus Mahogany	10.44 g	9.87 mL	
White Pine	2.56 g		0.37 $^g/_{mL}$

Summarizing Your Thoughts

8. In your own words, give definitions for the terms *base unit* and *derived unit*.

9. Look at the table of woods. Does the heaviest piece of wood have the greatest density? Why is it sometimes beneficial to describe a material's density rather than its weight or volume?

Team Skills

10. What was the hardest question for your group in this activity? What made it the most difficult for you to understand?

11. Gasoline is less dense than water. If you are given 100 g of gasoline and 100 g of water, which will have the greater volume, and why?

12. If the woods in the table above are each added to water, the red oak, balsa, and white pine float, while the eucalyptus mahogany and ironwood sink. From this information, give an estimate on the density of water, and describe how your group arrived at that estimate.

ACTIVITY 4.5 THE SIZE OF THE ATOM

Objective

- Through the use of SI units, gain greater appreciation of the nanoscale view of the atom

Getting Started

In order to make direct comparisons, you must ensure that all things being compared are in the same units. Now that you are capable of converting to nanometers, you can make some comparisons.

The Model

A sulfur atom with a radius of 0.1 nm is placed in the center of a softball.

Exploring the Model

1. Estimate the radius of a softball in centimeters and convert your answer to units of nanometers.

2. In grammatically correct sentences, explain each step of the calculation required to convert the softball's radius from cm to nm.

3. What difference (in powers of ten) is there between the units of the softball and the sulfur atom?

4. Describe a procedure that would allow you to determine the distance from the outer edge of the sulfur atom to the outer edge of the softball.

5. Calculate the distance between the edge of the sulfur atom and the surface of the softball.

6. The average distance between the earth and the moon is 384,403 kilometers. Express this distance in meters and nanometers.

Summarizing Your Thoughts

7. Wanting to know what you are learning in school, your 5-year-old brother asks you, "How small is a sulfur atom?" Write a short paragraph about the size of a sulfur atom in language that your little brother would understand (hint: use comparisons).

Team Skills

8. As a group, discuss which units are used to describe common distances, explain why SI units make comparing two distances much easier than English units (feet, yards, miles).

9. As a group, determine what information in today's activity is important to memorize. What information is more likely to be looked up?

End of Chapter Exercises

Activity 4.1

1. Indicate which of the following units would be most appropriate for measuring the property of the substance indicated

    a. The mass of a car: gallons or pounds

    b. The temperature of water: degrees Celsius or minutes

    c. The distance between cities: feet or grams

    d. How long a phone conversation lasts: days or miles

2. Indicate which of the following units would be most appropriate for measuring the property of the substance indicated

    a. The volume of a coffee mug: gallons or fluid ounces

    b. The weight of this book: ounces or tons

    c. The time for a snail to move 1 meter: seconds or minutes

    d. The height of a tree: inches or miles

3. Indicate which of the following would be the most reasonable value for the measurement indicated

    a. The weight of a paperclip: 0.01 pounds   1.0 pounds   100 pounds

    b. The distance to the moon: 25 miles   2500 miles   250000 miles

    c. The weight of food you eat in 1 day: 1 pound   10 pounds   100 pounds

    d. The time it takes to walk 10 miles: 100 seconds 1000 seconds 10000 seconds

    e. The temperature on Mars: -1000 F   0 F   1000 F

    f. The volume of a railroad boxcar: 60 gallons   6000 gallons   60000 gallons

Activity 4.2

4. There are 4 gills in 1 mutchkin, and 2 mutchkins in 1 chopin. Express this statement as an equality.

5. What is the conversion factor for gills to mutchkins?

6. What is the conversion factor for mutchkins to gills?

7. If you have 6 mutchkins, how many gills do you have?

8. There are 2 mutchkins in 1 chopin. If you have 12 gills, how many chopins do you have? Show your work and label all conversion factors.

9. Explicitly show how the conversion factors are used to calculate the number of seconds in 10 minutes.

10. If I give you 2 pounds of silver, how many ounces would you have?

11. If a B-52 bomber weights 35000 pounds, how many tons is this? (1 ton = 2000 lbs.)

12. Consider each of the following sets of units. If a measurement was taken in the first unit and converted to the second unit, would you expect the numerical part of the measurement to increase, decrease or stay the same?
    a. Pounds to ounces
    b. Seconds to minutes
    c. Miles to yards
    d. Gallons to cups
    e. Days to weeks
    f. Inches to feet
    g. Pints to quarts

13. Estimate without a calculator whether the number of seconds in 32 minutes is greater than or less than 500. What would you do if the answer shown on your calculator doesn't agree with your estimate?

14. Mark and Ken take a trip together, and Mark measures their distance travelled in miles, while Ken measures it in kilometers. They plot their data together, and come up with the plot shown below. On the plot, indicate which line corresponds to Mark's measurements, and which corresponds to Ken's. Using complete sentences and proper grammar, state how you arrived at your conclusion.

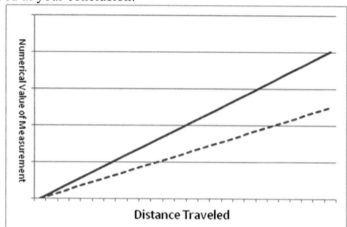

Activity 4.3

15. How is the SI unit for mass presented in Table 4.2 different from the other SI units?

16. How is converting between units of time different than converting between the other SI units?

17. Complete the following relationships between the numbers as expressed in standard notation and powers of ten by filling in the blanks (the first one is done for you)

$100 = 10^2$ _____ $= 10^7$

$0.1 =$ _____   _____ $= 10^{-5}$

_____ $= 10^{-1}$__ $100000 =$ _____

$1000000 =$ _____   _____ $= 10^{-2}$

_____ $= 10^4$__ $10 =$ _____

18. Complete each of the following equalitites by filling in the blanks with the correct value

    a. 1 dm = _____ meters

    b. 10 dm = _____ meters

    c. 0.3 kL = _____ Liters

    d. 200 mg = _____ grams

19. Complete each of the following equalitites by filling in the blanks with the correct unit

    a. 1 dm = 100 _____

    b. 10 dm = 1000 _____

    c. 10 dkm = 1000 _____

    d. 800 cL = 0.008 _____

    e. 1 cg = 100 _____

    f. 2 μm = 0.02 _____

    g. 40000 dkm = 4000 _____

    h. 100000000 ns = 10 _____

20. Complete each of the following equalitites by filling in the blanks with the correct unit

    a. 1 dm = 10^2 _____

    b. 10 dkL = 10^{-2} _____

    c. 10 dkm = 10^2 _____

    d. 100 mL = 10^5 _____

    e. 1 cg = 10^0 _____

    f. 1 μm = 10^{-3} _____

    g. 10^{-2} dkm = 10^1 _____

    h. 1000 ns = 10^{-6} _____

21. What prefixes are used to describe computer memory? Why are these prefixes appropriate, whereas most of those listed in the table would not be appropriate?

Activity 4.4

Show all calculations for the problems below.

22. A car travels 249 miles in 8.3 hours. What is the car's average speed?

23. If you consume 2300 calories per day in the food you eat, how many calories would you consume in one year?

24. What is the mass of a bowling ball with a density of 1.37 $^g/_{mL}$ and a volume of 4181 mL?

Activity 4.5

Show all calculations for the problems below

25. The distance to the sun is defined as one astronomical unit, 92,955,820.5 miles, and the diameter of the planet Earth at the equator is 7926.28 miles. How many planets the size of Earth could fit in between Earth and the sun?

26. Approximately how many times larger is the earth than your head? (1 mile = 5480 feet)

27. The diameter of a carbon atom is believed to be approximately 1.54 Angstroms (1 Angstrom = 1 x 10^{-10} meters, or 100 picometers). If I have a model of a carbon atom which is 1.5 centimeters in diameter, approximately how many times larger than an actual carbon atom is my model?

5

It's a Small, Small World (Atomic Theory)

Getting Started

How do we describe something we cannot see directly with our eyes? Very often we turn to drawings or models that are familiar to us. Our use of the sphere to represent atoms allows us to think of atoms as balls. This approach works for some observations, but we can't use it to explain all observations.

The atom was first theorized by the Greek philosopher Democritus. Even in his day the discussion of matter was disputed. Democritus and Aristotle argued over the description of matter. It is interesting to think of these two people discussing something they couldn't see. We respect their arguments now, but their discussions might not have been much different than the ones you might have with your friends. One significant difference would be that they had their discussion in public, and the result of their discussion was eventually written down. Science advances today through public debate and publication. Not much different really.

It took almost 2000 years before Democritus' ideas came back around. Dalton and Lavoisier separately studied matter and chemical reactions in the late 1700s. They proposed rules and definitions, tested these ideas, discussed them, and finally an updated view of the atom was proposed.

Still, people were asking, "What would an atom look like if we could see it with our eyes?" A research group led by J.J. Thomson started doing experiments to probe the atom further. Thomson, E. Goldstein, and Sir James Chadwick performed experiments for more than 35 years. The result of their work is the current "view" of the atom. In this chapter, we will probe their view in order to "see" an atom.

Chapter Goals

* Be able to state and give the location of the three subatomic particles

* Understand the relative distance between the nucleus and the electrons

* Learn the charge on each subatomic particle

* Be able to estimate the relative mass of each particle

* Understand which subatomic particles are added or removed when an ion is formed from an atom

* Recognize how our chemical symbols communicate the number of subatomic particles that have been added or removed

* Understand how mass and atomic numbers are represented with a chemical symbol

ACTIVITY 5.1 IDENTIFICATION OF THE SUBATOMIC PARTICLES

Objective

- Be able to state and give the location of the three subatomic particles

Getting Started

We represent atoms as spheres. Although this representation is useful most of the time, it does not explain fundamental differences in the ways elements interact with light or with each other. J.J. Thomson was the first scientist to discover a subatomic particle. Ten years later Thomson and Goldstein found a second subatomic particle, and about 20 years after that Chadwick found a third subatomic particle. Study the Model to gain the perspective of the atom that resulted from their experiments and discussion.

The Model

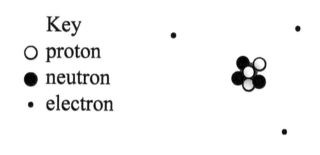

Key
O proton
● neutron
• electron

Figure 5.1   A Model of The Atom (Not to Scale)

Exploring the Model

1. How are the neutrons represented in the above model? _____

2. How many electrons are shown in the above model? _____

3. The nucleus is the term for the center of the atom, as shown in Figure 5.1. Which particles are present in the nucleus of an atom?

4. The nucleus is referred to as a dense part of the atom that contains most of the mass of an atom. What does *dense* part indicate about the masses of protons and neutrons as compared to electrons?

Summarizing Your Thoughts

5. Create a table that would be useful for remembering the names and relative location of the three subatomic particles.

Team Skills

6. In examining the Model, which particle would be expected to be easiest to remove from an atom and why?

ACTIVITY 5.2 RUTHERFORD'S EXPERIMENT

Objective

- Understand the relative distance between the nucleus and the electrons

Getting Started

As we noted in the last model, Figure 5.1 is not drawn to scale. The idea of the scale of the atom and where the particles are located puzzled many scientists for a long time until Ernest Rutherford (along with his students Hans Geiger and Ernest Marsden) ran his famous gold-foil experiment. Our model here is an oversimplification of Rutherford's experiment, but we can use it to gain some sense of the size of the nucleus with respect to the atom.

The Model

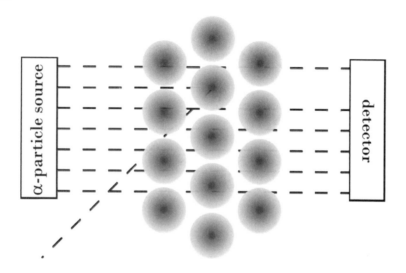

Figure 5.2   Simplified Rutherford Experiment

- α-particles are smaller than an atom, and positively charged

- Only approximately 1 in 20,000 α-particles are deflected.

- Spheres represent the atoms in the gold foil. We represent three layers of gold; Rutherford's foil would have had many more layers.

- (Alternate Model: lay ten coins on the table in an arrangement similar to the Model shown. Now think about how easy it should be to draw a line through the coins without disrupting their arrangement.)

Exploring the Model

1. From the information given in the text above, what percentage of the α-particles was deflected?

2. Rutherford estimated the diameter of the atom to be 10^{-8} cm. Estimate the maximum number of layers in his foil if the foil was 0.1 mm thick. (Hint: lay the spheres in a straight line with each edge touching.)

3. Speculate about why any particle would be deflected back toward the source. In the Model presented in Activity 5.1, the electrons are shown outside the nucleus. On the basis of how many particles hit the detector in Activity 5.2, is it reasonable to think that the α-particles collide with the electrons?

4. At the time of Rutherford's experiment, the prevailing thought about the structure of atoms was that they were solid spheres which were close together, touching one another to make up a solid object. How do the results of this experiment conflict with that view of the atom?

Summarizing Your Thoughts

5. Write a few sentences and provide a drawing that would explain why most of the α-particles passed through the foil, and include a brief explanation describing what caused the α-particles to be deflected.

Team Skills

6. Have each member of your group identify the most confusing part of this assignment. Then, as a group write a few sentences to clarify the confusion. Identify a resource that could be used to provide more clarity if the group cannot develop an acceptable explanation.

ACTIVITY 5.3 CHARGE AND MASS OF THE SUBATOMIC PARTICLES

Objectives

- Learn the charge on each subatomic particle

- Be able to estimate the relative mass of each particle

Getting Started

Now that we have identified the parts that make up an atom, it will be important for us to examine how changing the number of each particle changes two important properties: the charge and the mass. Review the Model to determine the fundamental charge and mass of the three particles.

The Model

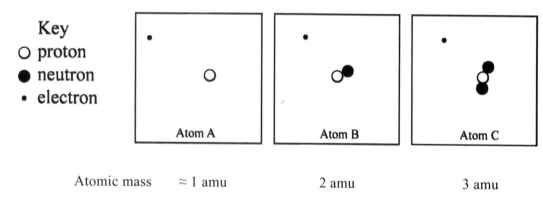

- The unit used for measuring the mass of an atom is the atomic mass unit (amu).

- The proton has a positive charge (+1). The overall charge for all three atoms in the Model is zero.

Exploring the Model

1. In atom A how many electrons are there? _____ In B? _____ In C? _____

2. In atom A how many protons are there? _____ In B? _____ In C? _____

3. In atom A how many neutrons are there? _____ In B? _____ In C? _____

4. If a proton has a mass of approximately 1 amu, estimate the mass of the electron. _____ amu

5. Estimate the mass of a neutron. _____ amu

6. What is the charge on the electron? _____

7. What is the charge on the neutron? _____

Summarizing your Thoughts

8. Recreate the table that you prepared in Activity 5.1, and add columns to include the
    approximate mass and charge of each sub-atomic particle. (The table should now include the
    name, charge, approximate mass, and the location of each subatomic particle.)

9. Atoms A, B, and C depicted in this activity are all *isotopes* of hydrogen. Based on your
    thoughts about the Model, write a sentence to describe the relationship between isotopes.

Team Skills

10. For your entire group there are five cookies available for participation. Using only whole
    cookies, divide these five cookies among your group members. The distribution of cookies
    should indicate how much each individual contributed to the completion of this activity.
    Write the name and number of points for each group member below.

ACTIVITY 5.4 PARTICLES RESPONSIBLE FOR THE FORMATION OF IONS

Objectives

- Understand which subatomic particles are added or removed when an ion is formed from an atom

- Recognize how our chemical symbols communicate the number of subatomic particles that have been added or removed

Getting Started

Though we will not get into all of it now, the addition or removal of one of the subatomic particles is important to the chemical reactions in which an atom can be involved. In later chapters we will look at electrons and how they affect reactions in much more detail. But for now, use the Model to learn which particles we are talking about, and how we symbolize the change in the number of these particles.

The Model

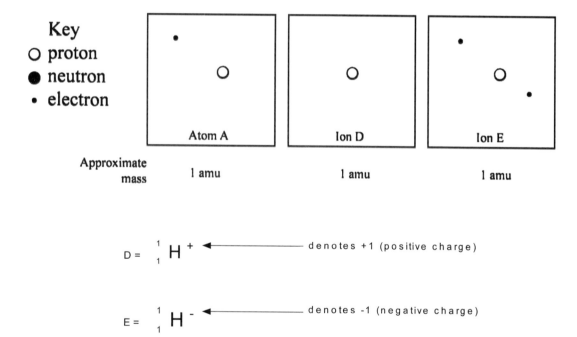

$$D = {}^{1}_{1}H^{+} \longleftarrow \text{denotes +1 (positive charge)}$$

$$E = {}^{1}_{1}H^{-} \longleftarrow \text{denotes -1 (negative charge)}$$

- Ion D is a cation, and Ion E is an anion

Exploring the Model

1. How many protons are in atom A? _____ In ion D? _____ In ion E? _____

2. How many electrons are in atom A? _____ In ion D? _____ In ion E? _____

3. How do the two chemical symbols differ for Ion D and Ion E?

4. One of the atoms is listed as an atom; the other two are listed as ions. What makes an ion different from an atom?

5. How does the atomic mass of the atom change when an electron is added or removed?

6. What is the charge on an ion when there are more electrons than protons? (+ or −) _____

7. What is the charge on an ion when there are more protons than electrons? (+ or −) _____

8. Indicate the correct term for this description: A cation/anion has more protons than electrons.

9. Indicate the correct term for this description: A cation/anion has more electrons than protons.

Summarizing Your Thoughts

10. Summarize what distinguishes an atom from an ion.

11. It sometimes surprises students that we use a − symbol to represent the ion where an electron has been *added*. Write yourself a note that will remind yourself when to use the + and − signs for ions.

Team Skills

12. Compared to the last group session how well did your group work together?

ACTIVITY 5.5 SYMBOLIC REPRESENTATIONS

Objective

- Understand how mass and atomic numbers are represented with a chemical symbol

Getting Started

We have seen throughout this book that symbols make up a large part of chemistry because without them it would be difficult to communicate all the information we gather experimentally. We will now continue with our study of the atom by taking a look at how we can represent symbolically all the information we have obtained about the number of each particle in the atom.

The Model

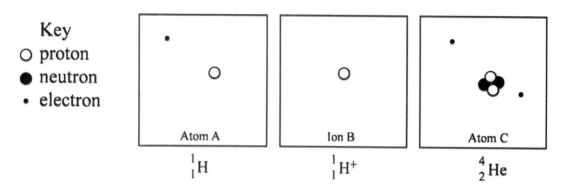

Key
○ proton
● neutron
• electron

Atom A   Ion B   Atom C

^1_1H $^1_1H^+$ ^4_2He

Exploring the Model

1. What elements are represented in the Model?

2. Which symbol represents the number of protons? (choose one)
   a. the superscript before the chemical symbol
   b. the subscript before the chemical symbol only
   c. the subscript before the chemical symbol and the chemical symbol
   d. the superscript after the chemical symbol

3. Describe the differences between the hydrogen atom and the helium atom in the Model.

4. Does the atom ^2_2H exist? What is it about this symbol that provides the answer?

5. What would be the chemical symbol (with atomic and mass numbers) for the atom presented in Figure 5.1? Check this answer with your instructor.

Summarizing Your Thoughts

6. A generic symbol for an element is shown. Write a key to this symbol that defines A, E, and Z.

$$^A_Z E$$

7. The mass number for the helium atom shown in the model is four. Review all the atoms shown in this activity, and write a definition for mass number.

8. What information around a chemical symbol changes with a changing number of neutrons?

9. What information around a chemical symbol changes with a changing number of electrons?

10. Given two chemical symbols with atomic number, mass number, and charge, how could you tell if the two particles represented were isotopes?

Team Skills

11. How well did your group stay on task through the course of the exercise? Describe steps you can take next time to keep everyone involved in the completion of the exercise.

End of Chapter Exercises

Activity 5.2

1. Rutherford estimated that the nucleus is 100,000 times smaller than the atom. Let's relate this estimate to a macroscopic relationship. Suppose that your favorite domed football stadium is one-half of an atom. The distance from the field surface to the dome is then the radius of the atom. Which ball would best represent the size of the nucleus: a marble, a golf ball, a soccer ball, or a beach ball?

2. Explain how you made your decision about the ball.

Activity 5.3

3. If it were possible to add another neutron to the atom C depicted in the Model, what would be its approximate atomic mass? _____

4. Estimate the atomic mass of an atom that contains 19 protons, 20 neutrons, and 18 electrons.
    _____

Activity 5.4

5. Classify each row in Table 5.1 as a cation, anion, or atom.

Table 5.1

Number of Neutrons	Number of Electrons	Number of Protons	Classification
9	10	9	Anion
13	12	12	
32	28	30	
12	10	11	
5	4	4	
20	18	19	
16	18	16	

6. In each of the following pairs of ions, which has more protons than electrons?

    a. Ca^{2+} or O^{2-}

    b. F^- or Fe^{3+}

    c. K^+ or I^-

Activity 5.5

7. What element or ion is represented by

    a. $^{37}_{17}E$ _____

    b. $^{195}_{78}E$ _____

    c. $^{23}_{11}E^+$ _____

8. Show how many protons, neutrons, and electrons are present for each species.

    a. $^{37}_{17}E$ protons: _____ neutrons: _____ electrons: _____

    b. $^{195}_{78}E$ protons: _____ neutrons: _____ electrons: _____

    c. $^{23}_{11}E^+$ protons: _____ neutrons: _____ electrons: _____

6

Significant Figures Are Not Just for History Class (Significant Figures and Scientific Notation)

Chapter Goals:

- Recognize how a reported value is related to the method of measurement

- Be able to define and identify measured numbers and exact numbers

- Be able to identify the number of significant figures in a measurement

- Distinguish between an exact number and a measured value

- Recognize the conventions of scientific notation

- Gain a sense of the magnitude of a value reported in scientific notation

- Determine how zeros are used when writing values in scientific notation

- Derive and write a set of rules for working with significant figures in addition and subtraction calculations

- Derive and write a set of rules for working with significant figures in multiplication and division calculations

- Determine the mathematical rules that define scientific notation operations

- Perform operations using numbers represented in scientific notation

ACTIVITY 6.1 REPORTING MEASURED VALUES

Objective

- Recognize how a reported value is related to the method of measurement

- Be able to define and identify measured numbers and exact numbers

- Be able to identify the number of significant figures in a measurement

Getting Started

When scientists look at a measured value, they will consider the value's accuracy, which is how close the value is to the accepted true value of a measurement. This thought process takes into account how the measurement was made and whether or not the measurement could be improved. For us all to understand each other, there must be some set of rules we can all understand. Following these rules is an ethical responsibility in order to communicate truthfully about how close data is to the actual value. Review this initial model to see how the reported value relates how a value was measured.

The Model

Model 1

Figure 6.1

Two students were asked to determine the number of beans in identical jars, with the same number of beans, shown in Figure 6.1. Each student took their jar home and came back to report.

- Anni: There are 584 beans in the jar

- Bart: There are 600 beans in the jar.

Model 2

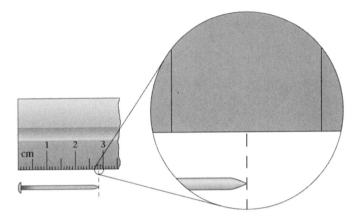

Figure 6.2

Anni and Bart are also asked to measure the length of the nail shown in Figure 6.2. Their reported measurements on the length of the nail are as follows.

- Anni: The nail is 2.8 cm long

- Bart: The nail is 2.85 cm long.

Exploring the Model

1. How is the reported value different between the number of beans Anni reported and the number of beans Bart reported?

2. How is the reported value different between the length of the nail that Anni reported and the length that Bart reported?

3. One of the students removed all the beans and counted them one at a time; the other one estimated the volume of a single bean to be about 1 cm^3, noted that the jar is labeled as 600 cm^3, and made an estimate. In the blanks provided, write the name of the student that adopted each procedure.

    a. Counted the beans: _____

    b. Estimated on the basis of size: _____

    c. Describe your reasons for making your assignments.

4. Which of the two measurement methods do you believe is the most accurate (which is closest to the actual number of beans in the jar)?

5. For the measurement of the length of the nail, whose measurement is more accurate?

6. For taking a measurement on an analog (non-digital) instrument, which method gives a more accurate result, rounding to the nearest mark, or estimating one digit past the markings on the instrument?

7. Anni's value for the number of beans in the jar is reported with three significant figures, while Bart's is reported with one significant figure. Anni's value for the length of the nail is reported to two significant figures, while Bart's is reported to three significant figures. Based on these statements, give a definition for the term *significant figure*.

8. In Bart's measurement of the nail, the last digit reported was the estimated digit. Based on this statement, give a definition of an *estimated digit*.

9. We can assume that for a good measurement, the actual value of the quantity measured would round to the reported value. For Anni's measurement, this would mean that the actual length of the nail would likely be between 2.75 cm and 2.85 cm, since numbers in that range would round to the reported value 2.8. What range for values of the length of the nail is suggested based on Bart's measurement?

Summarizing Your Thoughts

10. How can a reported number help you determine the most accurate measurement of a quantity?

Team Skills

11. As a group, establish a list of key terms and their definitions as explored in this activity.

ACTIVITY 6.2 EXACT NUMBERS AND MEASURED VALUES

Objective

- Distinguish between an exact number and a measured value.

The Model

Some statements including exact numbers are shown here:

      a. There are 584 beans in the jar.

      b. One foot may be divided into 12 inches.

      c. A car has four wheels.

      d. One millimeter is equal to 0.001 meters.

Some statements including measured numbers are shown below.

      e. I drove here at an average speed of 47.2 miles per hour.

      f. The container has a volume of 1.20 liters.

      g. The book weighs 3 pounds.

      h. The nail has a length of 3.54 centimeters.

Exploring the Model

1. Is there ever a circumstance when a foot would be 11 inches? (Einstein might differ with this one.)

2. If you carefully counted the beans in the jar three times, would you expect to get a different number each time you counted?

3. Consider how the nail was measured in Activity 6.1. Would it be possible for two people to report a different length of the nail (*i.e.,* Anni reports the length as 3.54 cm and Bart reports the length as 3.55 cm)?

Summarizing Your Thoughts

4. Using the model and your observations, give a definition for an *exact number*.

5. Using the model and your observations, give a definition for a *measured number*.

6. Why is it sometimes necessary to use methods that don't give an exact answer?

7. If a number is reported with 8 significant figures, how can the estimated digit be determined?

8. How can you tell if a reported number is an exact number or a measured number?

Team Skills

9. It is your group's responsibility to report the number of spiders in the forest. Even if all the members of your group went to the forest it is not possible to count all the spiders there, so you use a method that requires some approximations. Would it be ethical to report a value such as 1,632,388 spiders if you didn't count all of them? Why or why not?

ACTIVITY 6.3 SCIENTIFIC NOTATION AND MAGNITUDE

Objectives

- Recognize the conventions of scientific notation

- Gain a sense of the magnitude of a value reported in scientific notation

Getting Started

We deal with very large and very small numbers throughout chemistry. It is often difficult to work with values of these magnitudes, especially when we may know the value to only one or two significant figures. Thus we employ a notation that helps us write these values in a simpler manner. However, we need to make sure we can quickly grasp a value's magnitude when we look at it. This activity introduces the notation we use and explores how this notation should give us a sense of a value's magnitude. In the model below, a number is expressed in scientific notation and in standard notation, and more.

The Model

Scientific Notation:

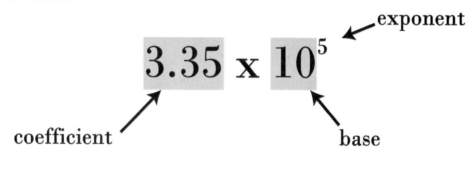

Standard Notation: 335000

Table 6.1

Standard Notation	Scientific Notation
3097	3.097×10^3
616	6.16×10^2
-59.67	-5.967×10^1
0.000138	1.38×10^{-4}
0.12	1.2×10^{-1}

Exploring the Model

1. How does the coefficient in scientific notation relate to the number as presented in standard notation?

2. How does the exponent in scientific notation relate to the number as presented in standard notation?

3. A negative number is presented in Table 6.1 above. What is different about how this number is represented in scientific notation?

4. Two of the numbers in Table 6.1 have a value between zero and one. What is different about how these numbers are reported in scientific notation?

Summarizing Your Thoughts

5. Why are some exponents positive and others negative?

6. Consider the two values, 3.67×10^{-3} and -3.67×10^{3}. What is the difference in meaning of the negative symbol ($-$) in these two values?

7. When writing a number in scientific notation, where do you place the decimal point in the coefficient?

8. For a number in scientific notation with a positive coefficient and negative exponent, where on the number line will the number be, $-\infty$ to -1, -1 to 0, 0 to 1, or 1 to ∞?

9. For a number that falls in the range between -1 to 0 on the number line, would you expect that number to have a positive coefficient and negative exponent, a negative exponent and positive coefficient, a positive coefficient and positive exponent, or a negative coefficient and negative exponent?

ACTIVITY 6.4 USING ZEROS IN SCIENTIFIC NOTATION

Objective

- Determine how zeros are used when writing values in scientific notation

Getting Started

Zeros cause a great deal of confusion when writing values in scientific notation. Sometimes they are significant, and other times they are simply indicating the magnitude of the number. Explore the Model to write a set of rules that helps you remember how to report zeros in a value.

The Model

Table 6.3

Standard Notation	Scientific Notation	Significant Figures
3097	3.097×10^3	4
−6001	-6.001×10^3	4
59.67	5.967×10^1	4
0.000128	1.28×10^{-4}	3
0.10	1.0×10^{-1}	2
−0.00504	-5.04×10^{-3}	3
600	6×10^2	1
600.	6.00×10^2	3

Exploring the Model

1. Which numbers in standard notation have zeros between two other numbers?

2. Which numbers in standard notation have zeros at the beginning of the value?

3. There are two values shown in standard notation for six hundred; how are they represented differently?

   Use the Model to help answer each question with a yes, no, or sometimes. If you decide to write "sometimes," provide an explanation of cases where you would answer "yes" or you would answer "no."

4. When a zero is between two numbers, is the zero included in the scientific notation?

5. When a zero is between two numbers, is it counted as a significant figure?

6.   When a zero is at the beginning of a value, is the zero included in the scientific notation?

7.   When a zero is at the beginning of a value, is it counted as a significant figure?

8.   When a zero is at the end of a value, is the zero included in the scientific notation?

9.   When a zero is at the end of a value, is it counted as a significant figure?

10. What is the relationship between the coefficient of a number in scientific notation and the number of significant figures?

11. Does it make a difference whether a value is positive or negative when counting the number of significant figures?

12. Ask your facilitator to look over your answers above, and initial here: _____

Summarizing Your Thoughts

13. What are some advantages of using scientific notation?

14. Work with your group to prepare a set of rules you could use later to convert a number presented in standard notation to a representation in scientific notation.

ACTIVITY 6.5 SIGNIFICANT FIGURES (ADDITION AND SUBTRACTION)

Objective

- Derive and write a set of rules for working with significant figures in addition and subtraction calculations

Getting Started

As you might suspect, as we perform calculations in chemistry we have to determine how to communicate the accuracy of a measurement that was made once the calculation has been completed. Estimations that are added to more accurate measurements need to be communicated. At the end of this exercise, you should be able to develop a rule for how to report a value to the proper number of significant figures after performing addition or subtraction on a set of values.

The Model 1

Figure 6.3 Identical jars of beans with the same number of beans in each jar.

Two students were asked to determine the number of beans in identical jars, with the same number of beans, shown in Figure 6.3. Each student took their jar home and came back to report.

- Anni: There are 584 beans in the jar
- Bart: There are 600 beans in the jar.

Exploring the Model

1. Anni's reported value is accurate to the hundreds, tens, or ones place?

2. Bart's reported value is accurate to the hundreds, tens, or ones place?

3. Restate how Anni and Bart determined these values in Activity 6.1. How is the method related to accuracy and the reported value?

4. Anni's jar of beans and Bart's jar of beans are emptied into a large bowl. Would it be accurate to describe the number of beans in the bowl as 1184 beans? Why or why not?

Model 2

We can assume that for a good measurement, the actual value of the quantity measured would round to the reported value. Consider the case of Anni and Bart's identical jars of beans being combined in a bowl. The total of the two reported values is 1184 beans, but the actual number (assuming that Anni counted the beans correctly) would be 1168 beans.

The minimum value of the range and the maximum value of the range are associated with rounding rules (Table 6.4). Using the example where the value is reported rounded to the tens place, any calculated value between 1175 and 1185 would round to 1180. With reporting the numbers to the ones place, the communicated accuracy represents that the actual number is within ½ bean, so that reported value represents and exact number of beans and high accuracy in counting.

Table 6.4

Last Place that is Reported	Reported Value	Minimum Value of Range	Maximum Value of Range
ones	1184	1183.5	1184.5
tens	1180	1175	1185
hundreds	1200	1150	1250
thousands	1000	500	1500

Exploring the Model

5. When Anni and Bart added their reported values together the sum is 1184, however the true number of beans is 1168. Which two reported values in Table 6.4 and their associated ranges would contain both the sum and the true number of beans?

6. Of the two reported values you selected, which reported value would communicate the *best* answer.

7. As a group, define *best* in the above question.

8. Review your responses to Model 1, compare the accuracy of the numbers reported by Anni and Bart. How are their reported values related to the best reported value you chose for the sum of their two containers?

Summarizing Your Thoughts

9. What trend or rule can help you determine how the accuracy of a measurement should be reported when separate measurements are added?

10. As a group, discuss whether or not your rule for addition should be applied to subtraction calculations. Describe your discussion, and provide an example that justifies your position.

11. Ask your facilitator to review your answers above, and initial here: _____

ACTIVITY 6.6 SIGNIFICANT FIGURES (MULTIPLICATION AND DIVISION)

Objective

* Derive and write a set of rules for working with significant figures in multiplication and division calculations

The Model

You go for a drive, and average 68.7 miles per hour on your trip over a period of 36 minutes. How accurately can the distance traveled be reported?

Exploring the Model

1. How many significant figures are present in the value provided for rate of travel?

2. How many significant figures are present in the value provided for time of the trip?

3. As stated in Activity 6.1, we can assume that for a good measurement, the actual value of the quantity measured would round to the reported value.

   For each of the values presented in the model, complete the table with the maximum and minimum values for rate of travel and time, as suggested by the number of significant figures for each value and the rules that are applied for rounding.

Property Measured	Reported Value	Minimum Value of Range	Maximum Value of Range
Speed	68.7 mph		
Time	36 minutes		

4. Let's now think about the range of how far you could have traveled on your trip. Complete the following calculations:

   a. For the minimum distance traveled:

      [minimum of the speed range] × [minimum of the time range]   =

   b. For the maximum distance traveled:

      [maximum of the speed range] × [maximum of the time range]   =

5. If we use the original values presented to calculate the distance traveled, the distance can be reported with one, two, three, or four significant figures. Complete the table below, reporting this value to each number of significant figures, and give the maximum and minimum values that are suggested with each number of significant figures.

Number of Significant Figures	Reported Distance	Minimum Value of Distance	Maximum Value of Distance
4			
3			
2			
1			

6. In comparing the minimum and maximum distance traveled that you calculated to the data in the table above, reporting the answer to how many significant figures most accurately reflects the possible range of distance traveled?

7. Compare the number of significant figures in the initial data provided in the model to your answer to question #6. What do you see in comparing these numbers?

Summarizing Your Thoughts

8. What trend or rule can help you determine how the accuracy of a measurement should be reported in calculations in which numbers are multiplied or divided?

9. Ask your facilitator to review your answers above, and initial here: _____

Team Skills

10. There are some cases in which this rule does not always apply. For example, a measurement of 2.723 feet is recorded. If the measurement is desired in inches, the properly reported result would be 32.676 inches (with 5 significant figures), although to reach this number we multiplied a number with 4 significant figures by a number with 2 significant figures (12 inches = 1 foot). Append your description above to describe why this is the case, using the terms measured numbers and exact numbers.

11. Contrast the rules that your group derived for reporting significant figures in addition and subtraction, and in multiplication and division. Give two differences between how significant figures are determined for each.

12. How would you determine how many significant figures to report for a calculation which involves both multiplication and addition. For example, if two measurements of 8.42 inches and 23.1 inches are to be added and converted to centimeters (1 cm = 0.3937008 inches), what should be the value reported? Give your value and how you arrived at the number of figures to report.

ACTIVITY 6.7 USING SCIENTIFIC NOTATION IN CALCULATIONS

Objectives

- Determine the mathematical rules that define scientific notation operations
- Perform operations using numbers represented in scientific notation

Getting Started

When large numbers are added, subtracted, multiplied, and divided, a large number of zeros or additional coefficients make these processes challenging. Using scientific notation can simplify these steps.

The Model

Table 6.5

Mathematical Operation	Normal Notation	Scientific Notation	Answer
Addition	5,000 + 1,000 =	$5.0 \times 10^3 + 1.0 \times 10^3 =$	6.0×10^3
Subtraction	5,000 − 1,000 =	$5.0 \times 10^3 − 1.0 \times 10^3 =$	4.0×10^3
Multiplication	5,000 x 1,000 =	$5.0 \times 10^3 \times 1.0 \times 10^3 =$	5.0×10^6
Division	5,000 / 1,000 =	$5.0 \times 10^3 / 1.0 \times 10^3 =$	5.0×10^0

Exploring the Model

1. What is 5,000 in scientific notation?

2. According to Table 6.5, what is the answer for 5,000 + 1,000 in scientific notation?

3. What is the answer for 5,000 x 1,000 in scientific notation?

4. In the value 5.0×10^0, what does it mean to multiply by 10^0?

5. In the Model, does the exponent in scientific notation change if it is added to a number containing the same scientific notation exponent? What if these numbers are subtracted?

6. What happens to the coefficient when it is added in scientific notation? What about subtracted?

7. Refer to the Model, and explain what happens to exponents if numbers in scientific notation are multiplied.

8. What happens to coefficients when they are multiplied in scientific notation? What if they are divided?

9. Refer to the Model, and explain what happens to exponents if numbers in scientific notation are divided.

Summarizing Your Thoughts

10. Describe how you can quickly estimate the magnitude of a value by using your observations about how the exponents change for the different mathematical operations.

End of Chapter Exercises

1. Consider measurements of two different objects, one is reported as 537 centimeters long, and the other is reported as 1.28 centimeters long. Comment on the number of significant figures of each, and the accuracy of each measurement.

2. Indicate in which range of the number line you would expect to find the value 5.2×10^3.(A, B, C, or D)

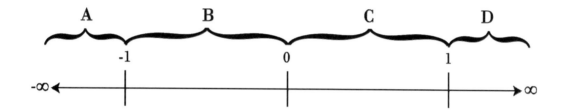

3. Indicate which range on the number line you would expect to find the value 5.2×10^{-3}.

4. Indicate which range on the number line you would expect to find the value -5.2×10^3.

5. Which number is larger, 2.3×10^3 or 8.4×10^2? Explain your choice.

6. Which number is larger, 8.4×10^{-4} or 4.7×10^{-1}? Explain your choice.

7. Which number is greater than one: 3.3×10^3 or 3.4×10^{-4}? Explain your answer.

8. Complete the following table:

Table 6.4

Value	Scientific Notation	Significant Figures
2014		4
	-7.20×10^5	
0.0094		
5964.20		
	8.2×10^3	
	2.841×10^{-4}	

9. Fill in the empty spaces in Table 6.6, reporting Answers to the correct number of significant figures:

Table 6.6

Mathematical Operation	Normal Notation	Scientific Notation	Answer
	$200.0 - 0.3 =$	$2.000 \times 10^2 - 3.0 \times 10^{-1} =$	
	$63 + 500. =$		5.63×10^2
Multiplication		$6.0 \times 10^5 \times 1.0 \times 10^{-3} =$	
	$7,200 / 0.001 =$	$7.20 \times 10^3 / 1.0 \times 10^{-3} =$	

10. If Anni weighed 135 pounds and Donnie weighed 175 pounds, how many total pounds would they weigh together? Show your calculation in scientific notation.

11. How can significant figures and scientific notation quickly help a reader assess how well we know a measured quantity?

12. If Anni predicted there were 5.82×10^2 beans in a jar, and Bart predicted that there were 6×10^2 beans, explain which of the two reported values is more accurate, and how that is communicated in the values given.

12. One string is measured and a length of 14.59 cm is recorded, and a second string is measured, and a length of 27.1 cm is recorded. What value should be reported for the sum of these two numbers?

13. If you fill up your 12 gallon gas tank in your car to capacity, and 0.04 gallons of gas is added as the gasoline drains out of the hose, what would be the reported value for the amount of gasoline held in your tank?

14. You have a piece of licorice with a length of 18.327 inches. If you cut 3.7 inches off of the end to give to a friend, what is the length of the remaining licorice? Report with the correct number of significant figures.

15. A rope bridge over a deep gorge states that it can support exactly 105 kilograms safely. You know that you weigh 52.3 kilograms. If your friend traveling with you states they weigh 50 kilograms, would it be safe to cross the bridge at the same time as them? Why or why not?

7

Periodic Advances at the Table (The Periodic Table)

Chapter Objectives

- Organize a group of elements by some commonality
- Relate the number of electrons available for bonding to the common charges

ACTIVITY 7.1 ORGANIZING THE ELEMENTS

Objective

- Organize a group of elements by some commonality

Getting Started

The story of how the modern periodic table was organized is quite amazing. At the time there were very few elements known, and everyone was trying to figure out some way to organize them in such a manner that you could predict the types of compounds and chemical reactions that would occur without placing yourself in danger. A very private Russian chemist named Dmitri Mendeleev proposed a simple, rational organization. Others at the time were offering a similar model, but Mendeleev made a few (at the time) outlandish predictions that came true. It is for the simplicity of his model and its ability to make good predictions that Mendeleev is most often credited with the modern periodic table. As we begin here, we won't reproduce Mendeleev's insight but rather try to see how we might start to group different elements together. Look at the Model to see how we might begin to organize the elements on the periodic table.

The Model

A simple anion: chloride, Cl^-

Some compounds containing the chloride anion: $NaCl$, $CaCl_2$, $LiCl$, $BaCl_2$, $MgCl_2$, KCl

A simple cation: sodium cation, Na^+

Some compounds containing the sodium cation: $NaCl$, Na_2O, $NaBr$, NaN_3, NaF, Na_2S

Exploring the Model

1. If the charge on the chloride ion is -1, then what is the charge on the lithium cation, $LiCl$? _____

2. If the charge on the sodium ion is $+1$, then what is the charge on the oxygen in Na_2O? _____

3. List all the elements (ions) from the Model separately, and group them according to their charge.

1 H Hydrogen																	2 He Helium
3 Li Lithium	4 Be Beryllium											5 B Boron	6 C Carbon	7 N Nitrogen	8 O Oxygen	9 F Fluorine	10 Ne Neon
11 Na Sodium	12 Mg Magnesium											13 Al Aluminium	14 Si Silicon	15 P Phosphorus	16 S Sulfur	17 Cl Chlorine	18 Ar Argon
19 K Potassium	20 Ca Calcium	21 Sc Scandium	22 Ti Titanium	23 V Vanadium	24 Cr Chromium	25 Mn Manganese	26 Fe Iron	27 Co Cobalt	28 Ni Nickel	29 Cu Copper	30 Zn Zinc	31 Ga Gallium	32 Ge Germanium	33 As Arsenic	34 Se Selenium	35 Br Bromine	36 Kr Krypton
37 Rb Rubidium	38 Sr Strontium	39 Y Yttrium	40 Zr Zirconium	41 Nb Niobium	42 Mo Molybdenum	43 Tc Technetium	44 Ru Ruthenium	45 Rh Rhodium	46 Pd Palladium	47 Ag Silver	48 Cd Cadmium	49 In Indium	50 Sn Tin	51 Sb Antimony	52 Te Tellurium	53 I Iodine	54 Xe Xenon
55 Cs Cesium	56 Ba Barium	57 La* Lanthanum	72 Hf Hafnium	73 Ta Tantalum	74 W Tungsten	75 Re Rhenium	76 Os Osmium	77 Ir Iridium	78 Pt Platinum	79 Au Gold	80 Hg Mercury	81 Tl Thallium	82 Pb Lead	83 Bi Bismuth	84 Po Polonium	85 At Astatine	86 Rn Radon
87 Fr Francium	88 Ra Radium	89 Ac** Actinium	104 Rf Rutherfordium	105 Db Dubnium	106 Sg Seaborgium	107 Bh Bohrium	108 Hs Hassium	109 Mt Meitnerium									

Figure 7.1   The Periodic Table, Main Group and Transition Elements

4. Circle the elements (ions) which are presented in the Model on the periodic table.

5. Does the table organize these elements according to your grouping based on the charge you determined from the list of compounds? Describe your observations thus far, and explain whether or not these initial observations help make sense of the Periodic Table.

Summarizing Your Thoughts

6. Write a sentence or two describing the relationship between an element's group (column) position and its charge.

7. Summarize all the ideas that your group developed about the organization of the periodic table. Highlight one idea that your group would like to know more about.

ACTIVITY 7.2 VALENCE ELECTRONS AND COMMON CHARGES

Objective

- Relate the number of electrons available for bonding to the common charges

Getting Started

Interestingly, if you look back at our last activity, you will see that we avoided the groups in the middle of the periodic table (scandium to zinc). We did this for a reason: predictions for those elements are a bit more complicated (and we'll leave that for a general chemistry class). Thus, let us introduce a little bit of nomenclature: main group elements and transition elements. We will be working with the main group elements (those shown in Figure 7.2) and will leave the transition elements for another discussion in some other setting.

Also notice from the last model that we grouped the elements by their common charges. That should be a clue that the electrons are important. Recall from earlier activities that cations and anions result from the loss or gain of an electron, respectively. Thus, it is time to think a little farther into how the electrons could be important to the organization of the periodic table.

But before we begin we need a little thought game. Electrons are a big part of why atoms hang out together. If that is true and there are many electrons around, then we need to ask ourselves, "Are some electrons more important than others when compounds form?" The answer is a definite "yes"!

As we group the atoms, we will group them according to the electrons that are farthest away from the nucleus because they will be the electrons available to do chemistry. The electrons doing the chemistry are called *valence electrons*. There is a quantum mechanical description explaining which electrons we should consider (think messy calculus), but a simple view is to think about the periodic nature of things.

The idea of valence was discussed extensively and the use of dots in the Model was initially described by G.N. Lewis. So the symbolism in the Model is often referred to as Lewis dot structures.

Use this activity to find the simple view and hope that as you progress in your studies you will add to your knowledge to gain a greater understanding.

The Model

Group 1A	Group 2A	Group 3A	Group 4A	Group 5A	Group 6A	Group 7A	Group 8A
H•							He:
Li•	•Be•	•Ḃ•	•Ċ•	:Ṅ•	:Ö•	:F̈•	:N̈e:
Na•	•Mg•	•Ȧl•	•Ṡi•	:Ṗ•	:S̈•	:C̈l•	:Är:
K•	•Ca•	•Ġa•	•Ġe•	:Äs•	:S̈e•	:B̈r•	:K̈r:

Figure 7.2   A Periodic Table Showing Only the Main Group Elements

The common cation of lithium is Li$^+$. The lithium cation has two electrons, the same number of electrons that the neutral (no charge) helium atom has.

The common ion of fluorine is F$^-$. The fluoride ion has the same number of electrons as the neutral (no charge) neon atom.

Exploring the Model

1. If a fluorine atom contains nine electrons, how many electrons does the fluoride ion possess?

2. How many total electrons are in a neon atom? _____

3. Describe the properties of the Group 8A elements. What do you know about these elements? Have you ever heard of chemical compounds of these elements?

4. For each of the Group 1A (a group is a column) elements, what is the expected charge? _____

5. For each of the Group 7A elements, what is the expected charge? _____

6. How many total electrons does potassium have when it forms its cation? _____

7. How many total electrons does chlorine have when it forms its anion? _____

8. How many total electrons does argon have? _____

9. How many dots are drawn around argon? _____

10. Let's let the number of dots be the available electrons. How do your observations about the ions of potassium and chlorine relate?

Summarizing Your Thoughts

11. Describe the relationship between the common ions and the number of electrons available to do chemistry (the number of dots we draw).

12. Write a couple of short sentences that will help you remember how the column position helps you remember how many dots to add or remove to determine the common ions.

13. Describe how your group completed this activity. Who led the discussion? Who recorded the answers? Did anyone disagree with the answers submitted?

Periodic Table of the Elements

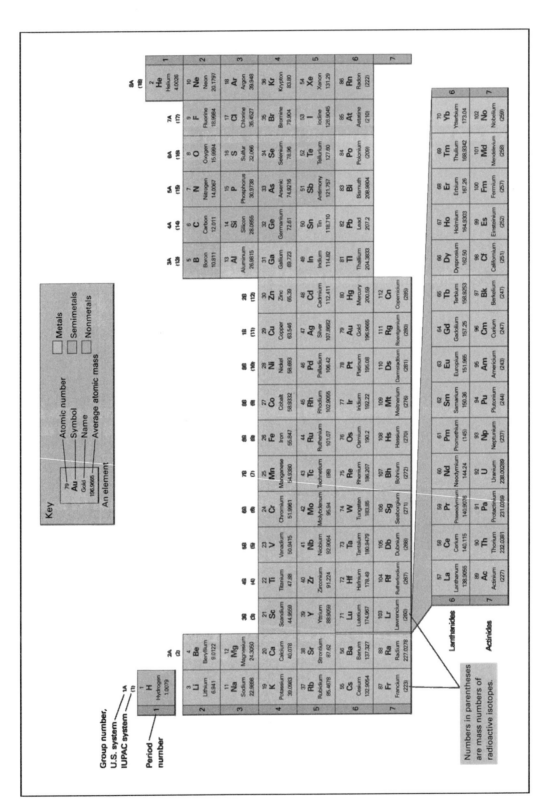

End of Chapter Exercises

Activity 7.1

1. Fluorine is found in nature as an ion; what charge is expected for the ion of fluorine? _____

2. What charge would you expect for the ion formed by strontium? _____

3. Predict the compound that would be expected from the common ions of calcium and oxygen.

4. Predict the compound that would be expected from the common ions of cesium and bromine.

Activity 7.2

5. For each dot structure, provide an example element, and then predict the common charge on that element:

:Ẍ: Element: _____   common ion of the element: _____

·X·   Element: _____   common ion of the element: _____

·Ẍ·   Element: _____   common ion of the element: _____

X·   Element: _____   common ion of the element: _____

8

A Compound By Any Other Name
(Classification of Ionic and Covalent
Compounds and Their Nomenclature)

Chapter Goals:

- Differentiate elements from compounds by molecular formula

- Know how the number of atoms of each element present in a compound are communicated in the chemical formula

- Be able to distinguish ionic and covalent compound based on the position of the component elements in the periodic table

- Identify some simple rules about ionic compound nomenclature (naming)

- Be able to use proper terms to describe cations and anions

- Be able to describe what happens when an atom ionizes

- Be able to predict the correct number of cations or anions in a binary ionic compound

- Understand how to write the chemical formula of ionic compounds containing metals with varying oxidation states

- Recognize the names of polyatomic ions, and understand how to write the chemical formulas of ionic compounds containing polyatomic ions

- Be able to determine the formula of an ionic compound containing a polyatomic ion

- Given a covalent compound's name, be able to give the proper chemical formula for the compound

- Given a covalent compound's chemical formula, be able to give the proper name for the compound

ACTIVITY 8.1 COMPOUNDS AND THEIR FORMULAS

Objectives

- Differentiate elements from compounds by molecular formula
- Know how the number of atoms of each element present in a compound are communicated in the chemical formula

Getting Started

Review the definitions for an element and a compound, review metals and nonmetals.

The *elemental form* of an element refers to how that element is most commonly found in nature in its pure form (not combined with other elements or as an ion). For most elements, we can think of them as individual atoms, and typically use their chemical symbols to represent them in chemical processes. Some elements most frequently occur not as individual atoms but as groups of atoms, so are symbolized differently.

The Model

Fe is the elemental form of iron.

C is the elemental form of carbon.

Cl_2 is the elemental form of chlorine.

$FeCl_3$ is a compound formed from the elements iron and chlorine.

CCl_4 is a compound formed from the elements carbon and chlorine.

Exploring the Model

1. What does the subscript 2 indicate in Cl_2?

2. Evaluate the statement, "The formulas for pure elements never contain a subscript." Is this statement true?

3. From the two examples provided, would you expect the formula S_8 to represent a compound or an element?

4. Using examples from the Model, explain how you classified S_8.

Summarizing Your Thoughts

5. What clues are given in a chemical formula that allow you to differentiate between an element and a compound?

6. As a group, discuss and describe the difference between an element and the elemental form of a substance.

7. The symbol O_2 can be described by many terms, including molecular oxygen, elemental oxygen, diatomic oxygen, and atmospheric oxygen. In the spaces below, describe what each of these terms communicates about O_2.

    Molecular oxygen -

    Elemental oxygen -

    Diatomic oxygen -

    Atmospheric oxygen -

ACTIVITY 8.2 RECOGNIZING IONIC AND COVALENT COMPOUNDS

Objective

- Be able to distinguish ionic and covalent compound based on the position of the component elements in the periodic table

Getting Started

There are two major classes of compounds typically encountered as part of an introductory course: ionic and covalent compounds. The concepts describing how these compounds are held together can be developed as you progress through your studies. However, before you get to those concepts you must be able to quickly classify a compound into one class or the other. In other words, your ability to classify compounds will guide how you will think about bigger ideas.

There are clues in the chemical formula. It is your job to use the Model to find these clues.

The Model

Table 8.1   Compounds That Are Considered...

Ionic	Covalent
$ZnCl_2$	CCl_4
Na_2O	P_2O_5
Fe_2O_3	CO
CuI	NI_3

Exploring the Model

1. Does the classification of whether compounds are ionic or covalent seem to be made based on how many atoms of each element are represented in the formula?

2. Write the symbols for the elements presented by the Model near their correct location on the outline of the periodic table.

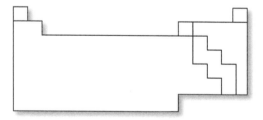

Figure 8.1   Outline of The Periodic Table, Main Group and Transition Elements

3. Compare the types of elements found (metals or nonmetals) for the two classes of compounds. What trends may be observed in the type of elements present and the classification of ionic or covalent compounds?

Summarizing Your Thoughts

4. Write a simple rule that will allow you to classify compounds as ionic or covalent on the basis of what you have learned from the Model.

Team Skills

As a group, develop consensus answers for the following questions:

5. What is the most difficult part of identifying whether a compound is ionic or covalent, based on the chemical formula of the compound?

6. Why is the periodic table a valuable tool for determining whether a compound is ionic or covalent?

ACTIVITY 8.3 NOMENCLATURE OF IONIC COMPOUNDS OF GROUPS I AND II

Objective

- Identify some simple rules about ionic compound nomenclature (naming)
- Be able to use proper terms to describe cations and anions
- Be able to describe what happens when an atom ionizes

The Model

Examine the table below, and answer the following questions.

Table 8.2

Cation	Anion	Chemical Formula	Compound Name
Na^+	Cl^-	NaCl	sodium chloride
Ca^{2+}	O^{2-}	CaO	calcium oxide
Zn^{2+}	Cl^-	$ZnCl_2$	zinc chloride
Li^+	S^{2-}	Li_2S	lithium sulfide
K^+	N^{3-}	K_3N	potassium nitride

Exploring the Model

1. Ionization is a term that describes how atoms or groups of atoms become charged. The sulfur atom gains two electrons when it ionizes to form the anion shown. What is the charge on the resulting sulfide anion?

2. When a potassium atom ionizes, how many electrons must be exchanged to result in the ion shown in the table, and are electrons gained or lost?

3. When the name of an ionic compound is given, which ion is stated first (the cation or the anion)?

4. Compare the first part of the compound names to the name of the element from the periodic table. How does the name of the cation correspond to the name of the element from which it comes?

5. Compare the second part of the compound name to the name of the element from the periodic table. How does the name of the anion correspond to the name of the element from which it comes?

6. From what part of the periodic table do the cations in the Model come (metals or nonmetals)?

7. From what part of the periodic table do the anions in the Model come?

Summarizing Your Thoughts

8. In what way did the name of each ion provide clues about the classification of each element as a cation or anion?

9. Based on the table presented, where on the periodic table would you expect to find elements that ionize to form cations?

10. Where on the periodic table would you expect to find elements that ionize to form anions?

11. Consider the clues you identified, and write a general rule for how you change the name of elements to cations when naming ionic compounds.

12. Consider the clues you identified, and write a general rule for how you change the name of elements to anions when naming ionic compounds.

13. Given the chemical formula of an ionic compound, list at *least* three necessary steps to give the correct name of that compound. (If needed, use a chemical formula of a compound from the table above as an example in listing the naming steps.)

Team Skills

As a group, develop consensus answers for the following questions:

14. In each ionic compound, are the number of ions in the compound communicated in the compound name? Give evidence from Table 8.2 to support your answer.

15. Describe how each of the following differs between an atom and the ion that is formed from that atom:

    a. the number of protons in an atom, and the number of protons in a cation formed from that atom

    b. the number of electrons in an atom, and the number of electrons in an anion formed from that atom

c. the number of neutrons in an atom, and the number of neutrons in an ion formed from that atom

d. the number of electrons in an atom, and the number of electrons in a cation formed from that atom

e. the mass of an atom, and the mass of an ion formed from that atom

ACTIVITY 8.4 PREDICTING THE FORMULA OF IONIC COMPOUNDS FROM THE CHARGE ON THE IONS

Objective

- Be able to predict the correct number of cations or anions in a binary ionic compound

The Model

Examine the table below, and answer the following questions

Table 8.3

Cation	Anion	Chemical Formula	Compound Name
Na^+	Cl^-	NaCl	sodium chloride
Zn^{2+}	Cl^-	$ZnCl_2$	zinc chloride
Na^+	S^{2-}	Na_2S	sodium sulfide
K^+	N^{3-}	K_3N	potassium nitride

Exploring the Model

1. Sodium chloride is NaCl, and zinc chloride is $ZnCl_2$. Why are there more chloride ions in the zinc compound?

2. Sodium chloride is NaCl, and sodium sulfide is Na_2S. Why are there more sodium ions in the sulfide compound?

Summarizing Your Thoughts

3. Explain how the number of chloride ions needed in aluminum chloride can be determined (the aluminum ion is represented as Al^{3+}).

4. From the table and the answers above, what do you know about the overall charge on ALL ionic compounds?

5. List at *least* three necessary steps to obtain the correct formula of any simple ionic compound when the compound's name is given (use the compounds in the table as a guide).

6. List at *least* three necessary steps to obtain the correct name of any simple ionic compound when the compound's formula is given (use the compounds in the table as a guide).

7. When will the number of cations and anions in the formula for an ionic compound be the same?

Team Skills

8. The oxidation state of the sodium ion is +1, and the oxidation state of the zinc ion is +2. As a group, come up with a definition for the term *oxidation state*.

9. List all the members of your group in the order in which they contributed to the successful completion of this activity.

10. What can you do to make sure that all group members understand the material presented?

11. Have the instructor check your answers above, and initial here: _____

ACTIVITY 8.5 NOMENCLATURE OF IONIC COMPOUNDS OF THE TRANSITION METALS

Objective

*   Understand how to write the chemical formula of ionic compounds containing metals with varying oxidation states

Getting Started

When a Group 1A metal forms a cation, it will always form a +1 cation. When a Group 2A metal forms a cation, it will always form a +2 cation. However, as we progress into the transition metals we find that these metals can form cations with different charges under different circumstances. Use the Model below to develop some rules that describe how to communicate the charge of the cation.

The Model

Examine the table below, and answer the following questions

Table 8.4

Chemical Formula	Compound Name
$FeBr_2$	iron(II) bromide
$FeBr_3$	iron(III) bromide
PbO	lead(II) oxide
PbO_2	lead(IV) oxide
Cu_3N	copper(I) nitride
Cu_3N_2	copper(II) nitride

Exploring the Model

1. What is the expected charge on the bromide ion?

2. What is different about the *chemical formulas* of the last two compounds in Table 8.4 (Cu_3N and Cu_3N_2)?

3. What is different about the *compound names* of these last two compounds in Table 8.4 (copper (I) nitride and copper (II) nitride) ?

4. Use your rules developed in Activity 8.4 to determine the charge on the iron ion in these compounds:

    a. Charge on iron in $FeBr_2$:

    b. Charge on iron in $FeBr_3$:

5. How is the Roman numeral in the compound name related to the charge on the iron ions?

6. Does this hold true for all the compounds in the table above?

7. What types of metals require the use of a Roman numeral in the name of their ionic compounds?

8. Where are these metals located on the periodic table?

Summarizing Your Thoughts

Use complete sentences to answer the following questions

9. Why do the compounds in this activity require Roman numerals in the name while compounds such as calcium chloride do not?

10. If only the chemical formulas were given for the compounds in the above examples, how could you determine the amount of charge on the cation?

11. How will you know when to use a Roman numeral when writing the name of an ionic compound?

12. When naming an ionic compound which contains a metal cation with more than one possible oxidation state, how do the naming rules for that compound differ from one which contains a metal cation with only one possible oxidation state? Answer with complete sentences and proper grammar. (If needed, look at your answer from Activity 8.4, listing the steps necessary to give the correct name of an ionic compound from its chemical formula.)

13. Look at your answer from Activity 8.4 that lists the steps necessary to give the correct chemical formula of an ionic compound given its name. How do the steps differ when ions with varying charges are involved?

ACTIVITY 8.6 NOMENCLATURE OF THE POLYATOMIC IONS

Objective:

- Recognize the names of polyatomic ions, and understand how to write the chemical formulas of ionic compounds containing polyatomic ions

Getting Started

Thus far we have considered only simple, monoatomic cations and anions. There is another class of ions that are often called polyatomic ions. Polyatomic ions are a group of atoms that are held together by covalent interactions, and the entire group of atoms carries the charge. The most common polyatomic ions contain oxygen. Their names may not seem to make sense now, but there is a system to this madness. It is your task to study the Model and determine what the nomenclature rules are.

The Model

Table 8.6

Ion	Name	Ion	Name
N^{3-}	nitride	S^{2-}	sulfide
NO_2^-	nitrite	SO_3^{2-}	sulfite
NO_3^-	nitrate	SO_4^{2-}	sulfate

Exploring the Model

1. What element is associated with the prefix "nitr-"?

2. Does the suffix of each name depend on the charge of the ion?

3. Does the suffix tell you how many oxygen atoms there are?

4. Compare nitrate to nitrite. Which ion has more oxygen atoms?

5. Compare sulfate to sulfite. Which ion has more oxygen atoms?

Exercising Your Knowledge

6. Consider the two oxo- ions of chlorine, ClO_2^- and ClO_3^-. Which ion would have the –ate ending?

7. Consider the two oxo- ions of phosphorus, PO_3^{2-} and PO_4^{3-}. Which ion would have an –ate ending?

8. Write the names of these two oxo- ions of phosphorus.

Summarizing Your Thoughts

9. How does the name of a polyatomic anion differ from the name of a monoatomic anion?

Team Skills

10. The last three letters of a name can tell a lot about a particle! For each of the name endings
    below, give a general description of what type of ion or particle would be expected to have
    that ending (cation, monatomic anion, polyatomic anion, metal element, and/or nonmetal
    element).

   a.   –ide ion   _____

   b.   –ium   _____

   c.   –ate ion   _____

   d.   –ine   _____

   e.   –ite ion   _____

   f.   –ium ion   _____

11. Have your facilitator check your answers above, and initial here: _____

ACTIVITY 8.7 PREDICTING THE CHEMICAL FORMULA OF IONIC COMPOUNDS CONTAINING POLYATOMIC IONS

Objective

- Be able to determine the formula of an ionic compound containing a polyatomic ion

Getting Started

Now that we have introduced polyatomic ions, we have to consider how this new twist affects the name of a compound and how we write the chemical formula. There are new features in the Model; let's see if we can figure out how to handle them.

The Model

Table 8.7

Chemical Formula	Compound Name
$CaSO_4$	calcium sulfate
$CaSO_3$	calcium sulfite
Na_3PO_4	sodium phosphate
Li_2CO_3	lithium carbonate
NH_4Cl	ammonium chloride
$Be(NO_2)_2$	beryllium nitrite
$Mg_3(PO_3)_2$	magnesium phosphite
$Fe(NO_3)_3$	iron(III) nitrate
$Al(OH)_3$	aluminum hydroxide

Exploring the Model

1. The ammonium cation is the only polyatomic cation in the Model. What is the formula and charge of the ammonium cation?

2. Do all polyatomic ions require the use of parentheses in a chemical formula?

3. When are parentheses used?

4. Have the nomenclature rules you established earlier in this chapter changed? If so, how? If not, is that important to know?

5. How many nitrogen atoms are in beryllium nitrite? _____

6. Describe the thought process you used to determine the number of nitrogen atoms in beryllium nitrite.

7. How many oxygen atoms are in beryllium nitrite?   _____

8. In what way does determining the number of oxygen atoms differ from the process you just described for nitrogen?

9. If the parentheses were omitted and aluminum hydroxide was written as $AlOH_3$, how would that change the number of atoms of each element represented in the chemical formula?

10. How many of each element is present in aluminum hydroxide, as presented in the Model?

    g. Aluminum: _____

    h. Oxygen: _____

    i. Hydrogen: _____

Summarizing Your Thoughts

11. Write a rule that can be used to determine whether or not parentheses are needed when writing a chemical formula.

12. Explain why the number of each ion is not included in the name of an ionic compound.

Team Skills

13. Make a list of information you must know in order to write the correct formula for an ionic compound (this list may require reviewing all the activities in this chapter).

ACTIVITY 8.8 NOMENCLATURE OF COVALENT COMPOUNDS

Objectives

- Given a covalent compound's name, be able to give the proper chemical formula for the compound

- Given a covalent compound's chemical formula, be able to give the proper name for the compound

Getting Started

We will be using gases and other compounds as illustrations of naming covalent compounds. Covalent compounds are defined as groups of atoms that stay together because of shared electrons in chemical bonds. There are an infinite number of covalent compounds. Here, we will be focusing on naming some of the smaller covalent compounds.

The names of covalent compounds are similar to those of the ionic compounds, but there are differences. Use the Model to see if you can figure out how the rules differ.

The Model

Table 8.9

Compound Name	Compound Molecular Formula
phosphorus hexafluoride	PF_6
tetracarbon decahydride	C_4H_{10}
boron trichloride	BF_3
dinitrogen oxide	N_2O
carbon monoxide	CO
dinitrogen tetroxide	N_2O_4

Exploring the Model

1. Where on the periodic table do you find all the elements used in the Model (metals or nonmetals)?

2. What suffix is used for all the compounds? _____

3. How many atoms of nitrogen are present in dinitrogen tetroxide? _____

4. Use the Model to fill the following table with prefixes used to designate the number of each type of atom in a binary compound:

Table 8.10

Prefix	Number of atoms of element
(no prefix)	
	One
	Two
	Three
	Four
	Five
	Six
Hepta-	Seven
	Eight
	Nine
	Ten

Summarizing Your Thoughts

5. In the biological process called respiration, we inhale oxygen and exhale *carbon dioxide*. When fossil fuels are burned, a toxic gas that may be produced is *carbon monoxide*. Explain why you wouldn't use the name "carbon oxide" for these molecules.

6. In complete sentences state the rules for naming a covalent compound, given the compound's molecular formula.

Team Skills

7. Compare your rules for naming covalent compounds with the rules you established for writing the names of ionic compounds. As a team, come up with a consensus approach to determining how to name a compound given its chemical formula. Make sure your rules clearly help you decide when you use the prefixes indicating the number of atoms and when you use Roman numerals.

End of Chapter Exercises

1. What are three major ideas you learned in this chapter?

2. What new information did you learn in this chapter that adds to previous knowledge that you had about a topic?

Activity 8.1

3. Classify each formula below as a the elemental form of a metal, elemental form of a nonmetal, or compound.

Co   _____

$CaCl_2$ _____

CsOH   _____

Br_2 _____

NaBr   _____

SiO_2 _____

PF_5 _____

P_4 _____

OF_2 _____

4. Match each of the statements below with the molecular formula to the right that best matches the statement.

_____ Contains the most fluorine atoms   A. Xe

_____ Contains the most atoms   B. NaF

_____ Composed of metal and nonmetal elements   C. NF_3

_____ Represents one atom   D. F_2

                                                                E. $SiBr_4$

Activity 8.2

5. Classify each of the following as either ionic or covalent.

   NaBr   _____

   SF_6 _____

   $CoBr_2$ _____

   OF_2 _____

   NO_2 _____

   BaS   _____

   CsF_2 _____

   $CrCl_3$ _____

   CO_2 _____

   N_2O_4 _____

Activity 8.3

6. For each of the following, predict whether the ion will likely be a cation or an anion.
   a. magnesium ion
   b. selenide ion
   c. bromide ion
   d. cesium ion

7. For each ionic compound, identify the cation and the anion.
   e. sodium fluoride
   f. strontium sulfide
   g. lithium iodide
   h. barium chloride

Activity 8.4

8. How many chloride ions would combine with an Al^{3+} ion to form aluminum chloride?

9. What charge does the barium ion possess in the compound $BaCl_2$?

Activity 8.5

10. Complete the table that follows with the proper ions, chemical formulas, and compound names. The first row has been completed as an example.

Table 8.5

Cation	Anion	Chemical Formula	Compound Name
Na^+	Cl^-	NaCl	sodium chloride
Ba^{2+}	I^-	BaI_2	
Mn^{2+}	O^{2-}		manganese(II) oxide
Mg^{2+}	N^{3-}		
			cobalt(III) fluoride
		CrO	
Cu^+	S^{2-}		
		Ca_3P_2	
		SnS_2	

Activity 8.7

11. Complete the table below with the proper ions, chemical formulas, and compound names. The first row is completed as an example.

Cation	Anion	Chemical Formula	Compound Name
Na^+	Cl^-	NaCl	sodium chloride
		LiCN	lithium cyanide
Ca^{2+}	OH^-		
Fe^{2+}	NO_3^-		
			barium phosphate
Cr^{2+}	PO_3^{3-}		
K^+	SO_3^{3-}		
			ammonium carbonate
		$AuPO_3$	
			copper(II) cyanide

Activity 8.8

12. From the information given above, complete the following table.

Table 8.11

Compound Name	Compound Molecular Formula
sulfur difluoride	
	PCl_3
silicon dioxide	
	H_2S
carbon tetraiodide	
	$SiBr_2$
	P_4O_{10}

13. A dentist calls you up and needs to order more laughing gas for his dental clinic. You check in a chemistry reference and find that the chemical name for laughing gas is dinitrogen monoxide. You may order N_2O, NO, or NO_2. One is the correct compound and the other two are toxic gases. Which should be ordered to keep the patients happy and alive?

9

Counting To a Trillion Trillions (The Mole Concept)

Chapter Goals:

- Understand the relationship between the mass of an element and the number of particles (the mole)

- Write equalities, and use these equalities to create conversion factors

- Describe the how large the mole is compared to everyday quantities

ACTIVITY 9.1 HOW TO COUNT ATOMS

Objective

- Understand the relationship between the mass of an element and the number of particles (the mole)

- Write equalities, and use these equalities to create conversion factors

The Model

Beaker 1

55.85 g of iron

1.000 mole of iron

6.02×10^{23} atoms of iron

Beaker 2

111.6 g of iron

2.000 mole of iron

12.04×10^{23} atoms of iron

Beaker 3

112.0 g of cadmium

1.000 mole of cadmium

6.02×10^{23} atoms of cadmium

Exploring the Model

1. If the equality for Beaker 1 is 55.85 g of iron = 1.000 mole of iron, what is the equality for Beaker 2?

2. Using the equality 55.85 g of iron = 1.000 mole of iron, we can write the conversions factors

$$\frac{55.85 \text{ g of iron}}{1.000 \text{ mole of iron}} = 1 \quad and \quad \frac{1.000 \text{ mole of iron}}{55.85 \text{ g of iron}} = 1$$

Describe the mathematical operation required to write these conversion factors from the equality.

3. Write the conversion factors showing the relationship between grams and moles of iron for Beaker 2.

4. How are the conversion factors for Beakers 1 and 2 related?

5. Write the equality between moles of iron and atoms of iron.

6. Write the equality between moles of cadmium and atoms of cadmium.

7. Mathematically, show the relationships (the conversion factors) which can be used to demonstrate that although the contents of Beakers 1 and 3 have different masses, they contain the same number of atoms.

8. Use conversion factors to show that the relationship between moles and atoms is the same for all three beakers.

9. Compare the information in the Model for each element to the information provided on a periodic table. How is the mass (grams) related to the moles? How this relationship communicated numerically on the Periodic Table?

Summarizing Your Thoughts

10. Based on your observations from the activity, explain how the number of atoms may be calculated from the number of moles of an element. Include in your explanation the information you might need to obtain from the periodic table.

11. Explain the benefit of thinking in terms of moles when working with chemical quantities.

Team Skills

12. In complete sentences, please describe the steps your group has taken to work through the problems more quickly.

13. Is everyone in your group contributing equally? Explain how you ensured that everyone was contributing. What could you do next time to improve?

ACTIVITY 9.2 A MATTER OF SCALE

Objective

- Describe the how large the mole is compared to everyday quantities

The Model

$10^0 = 1$

$10^1 = 10$

$10^2 = 100$

$10^3 = 1\ 000$

$10^4 = 10\ 000$

10^1 is one power of 10 larger than 10^0

10^3 is two powers of 10 larger than 10^1

In other words, 1000 is 100 times larger than 10

The mole is defined as 6.02×10^{23} particles. Particles can be anything you can count as a single unit. Some examples:

- If you have a mole of apples, you have 6.02×10^{23} apples.
- If you have 6.02×10^{23} grains of sand on the beach, you have one mole of grains of sand.
- Given 6.02×10^{23} atoms of copper, you have one mole of copper.

Exploring the Model

1. How many more powers of 10 is 10^5 than 10^2?

2. Write the relationship between the mole and the number of particles as an equality.

3. Write the two conversion factors that could be written from this equality.

4. Write all the zeros required to take 6.02×10^{23} out of scientific notation and represent this number in standard notation.

5. Estimate the volume of a single grain of sand, and comment on how large a beach that contains one mole of sand would be.

   Our volume estimate for a grain of sand: _____ mL (hint: $1 \text{ cm}^3 = 1$ mL)

   Our calculated volume for a mole of sand: _____ mL

   Comments:

6. Compare your estimate for the size of the beach to the volume that you estimate for one mole of copper. Estimate how many factors of 10 the size of a single copper atom differs from a single grain of sand? (*hint*: use the density of copper as 8.94 g/mL)

   Our volume estimate for a mole of copper: _____ mL

Summarizing Your Thoughts

7. Explain in a few short sentences how to mathematically convert from moles of atoms to the number of atoms.

8. The magnitude of the mole is pretty big. Explain just how big this number is in language that could be understood by a small child.

Team Skills

9. Describe which part of this activity your group had the most difficulty and how your group addressed this difficulty.

10. Describe an important insight related to the magnitude of the mole that one member of your group gained from this activity.

End of Chapter Exercises

Activity 9.1

1. Samples of different masses and numbers of particles for several elements are listed below. Complete the Table with the missing information for each element.

Table 9.1

Element	Mass of Sample	Number of Particles in Sample	Number of Moles in Sample
Magnesium		6.02×10^{23} atoms	1.00 mole
Arsenic	150 grams		
	23.0 grams	6.02×10^{23} atoms	
Lithium	13.9 grams		
	34.3 grams		0.25 moles
Boron		3.01×10^{23} atoms	
Silicon	56.2 grams		
	40.4 grams	12.04×10^{23} atoms	
Iodine			0.25 moles
	100 grams		0.50 moles

2. If you have two samples, each with an actual amount of 100 g of silver and 100 g of gold, which sample has more atoms (or are they the same)? Explain your answer.

3. Based on how you thought about which had more atoms, 100 g of silver or 100 g of gold, explain why it is difficult to compare chemical quantities using the mass of a sample.

4. You have half a mole of nuts and three-quarters of a mole of bolts. Which sample has more pieces (or are they the same)? Show a set of calculations that supports your answer.

5. Complete the drawing by filling in the second beaker:

    Beaker 1   **Beaker 2**
    55.8 g of iron   77.2 g of iron

6. Explain how you made your estimate for the second beaker above.

7. Diamonds are made of carbon. Estimate the mass of a diamond set in a typical piece of jewelry, and calculate the number of moles of carbon and the number of atoms in the diamond (1 carat = 200 mg).

8. One mole of copper atoms would have a mass of about 63.5 grams. Look at the Model in Activity 9.1, and sketch in the beaker below how full you would expect it to be with one mole of copper.

9. If you have 3.01 x 10^{23} apples, how many moles of apples do you have?

10. How many atoms do you have if you have 0.5 mole of atoms?

11. Explain whether or not 0.5 moles of carbon atoms would have the same mass as 0.5 moles of silicon atoms. Why are they the same or why are they different?

12. How many moles do you need to have 6,020,000 atoms?

13. Explain whether or not it matters if the problem above is 6,020,000 atoms of carbon or 6,020,000 atoms of platinum.

Activity 9.2

14. Compare the number of ^{1}H atoms needed for a mass of 1 gram with the number of average-sized paper clips needed to have a mass of 1 gram.

15. How many more powers of ten atoms are there in 1 gram of ^{1}H than there are seconds in 13.7 billion years? (Recall that 10^3 is two powers of 10 larger than 10^1.)

16. Are the there more or fewer seconds in 13.7 billion years than in 1 mole of seconds?

17. By how many powers of 10 does 13.7 billion years in seconds differ from 1 mole of seconds?

18. Could a chemist live long enough to individually count the number of gold atoms in a ring?

10

It's a Balancing Act

Objectives

- Be able to balance a chemical equation

- State the Law of Conservation of Mass and how it applies to a balanced chemical equation

- Be able to identify products and reactants in a chemical equation

- Be able to represent a written description of a chemical reaction as a chemical equation

- Be able to balance a chemical equation

- Be able to use a balanced chemical equation communicate information about the quantities of compounds involved in a reaction

- Be able to use a balanced chemical equation to communicate mass quantities of reactants consumed or products formed in a chemical reaction

Periodic Table of the Elements

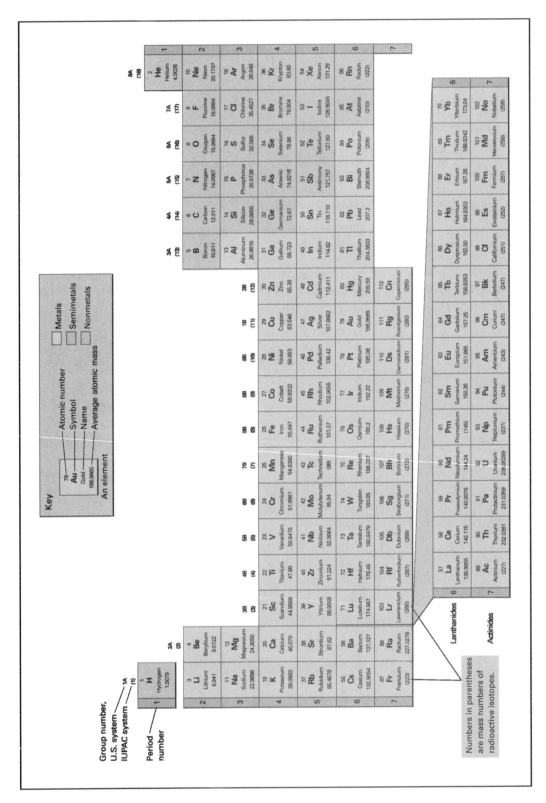

ACTIVITY 10.1 CHEMICAL EQUATIONS

Objectives

- Be able to balance a chemical equation

- State the Law of Conservation of Mass and how it applies to a balanced chemical equation

- Be able to identify products and reactants in a chemical equation

The Model

In studying for a chemistry exam, Ted becomes a bit hungry and decides to make a grilled cheese sandwich for a snack. Since chemistry is on his mind, Ted begins to think about this as a combination of bread and a slice of cheese, which may be represented as:

$$2\ Bd + Ch\ \rightarrow\ Bd_2Ch$$

Exploring the Model

1. Does the equation sufficiently describe the individual parts needed to make a sandwich?

2. What does the coefficient represent in front of the symbol representing bread?

3. How does a grilled cheese sandwich differ from two slices of bread and a slice of cheese?

4. In the morning, Ted makes a his super breakfast sandwich of two slices of bread, a fried egg, two slices of ham, and a toaster pastry (which come in packs of two). An equation representing this masterpiece is below. Put whole number coefficients in place to represent all the reactants being used up to make a whole number of sandwiches.

    _____ Bd + _____ Eg + _____ Hm + _____ Tp_2 → _____ Bd_2EgHm_2Tp

5. How many sandwiches does the equation describe making?

6. How would the equation above change if you needed to make four sandwiches? Does the equation need to change, or could it stay the same?

7. Going back to the original grilled cheese sandwich, if a slice of bread has a mass of 25 grams, and a slice of cheese has a mass of 20 grams, what is the mass of the grilled cheese sandwich?

8. How did you figure out this answer?

Summarizing Your Thoughts

9. The Law of Conservation of Mass may be stated as *in a chemical reaction mass is not created or destroyed, but instead changes form.* How is this demonstrated in making a grilled cheese sandwich?

10. In a balanced chemical equation what must be equal on both sides of the equation, the sum of the coefficients, the number of molecules, or the number of atoms of each element?

11. Why does a balanced chemical equation obey the Law of Conservation of Mass?

12. As a group, discuss and come up with an explanation of how a coefficient in front of a chemical formula or chemical symbol in a chemical equation differs from a subscript in a chemical formula.

13. In the Model, bread and cheese are reactants, and a grilled cheese sandwich is the product. Using these descriptions, give definitions for reactants and products in a chemical reaction.

ACTIVITY 10.2 VISUALIZING CHEMICAL REACTIONS

Objectives

- Be able to represent a written description of a chemical reaction as a chemical equation

- Be able to balance a chemical equation

Getting Started

We ask that you find some macroscopic objects (a marker) to represent atoms. Your macroscopic items could be small bingo chips, candies (M&Ms™ or Skittles™), or even wadded up pieces of paper. However, what is important as you begin is that you prepare a color-code key so that you can remember which atoms are being represented.

Suggested Marker Color Scheme		Color Scheme Used
Oxygen:	red	Oxygen:
Sodium:	purple	Sodium:
Chlorine:	green	Chlorine:
Calcium:	orange	Calcium:

The Model

Chemical reactions may be described in words or in a *chemical equation*, which is a representation of a reaction using chemical symbols and chemical formulas. One chemical reaction could be described in words as calcium oxide reacts with sodium chloride in a double-replacement reaction to form calcium chloride and sodium oxide.

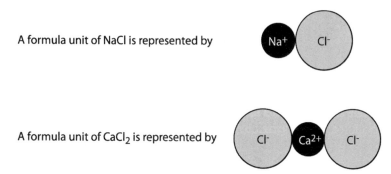

A formula unit of NaCl is represented by

A formula unit of $CaCl_2$ is represented by

Exploring the Model

1. What are the products in the chemical reaction written in the model?

2. To represent a reaction as a chemical equation, each component must be represented as a chemical formula. Fill in the blanks below with the proper chemical formula for each compound involved in the model reaction. Be sure to use the proper number of anions and cations to make the overall charge of each compound zero.

_____ + _____ → _____ + _____

3. Model this reaction using your macroscopic objects representing atoms. Start with *one formula unit of each* of the reactants in the proper grouping, and rearrange the objects to represent the product compounds formed.

    a. Were you able to form all the products indicated in the chemical equation?

    b. If yes, how were you able to do it? If not, why were you not able to complete the task, and what else is needed to form the products indicated?

4. In a chemical equation, coefficients are numbers in front of chemical formulas that indicate how many particles of each compound are required for the reaction to take place as written. In the blanks below, place the chemical formulas for each component in the chemical equation, and indicate before each chemical formula the number of particles (or groups) needed for the reaction to take place (the first one is done for you)

    $\underline{1}$ $\underline{CaO}$ + $\underline{\hphantom{xx}}$ $\underline{NaCl}$ $\rightarrow$ $\underline{\hphantom{xx}}$ $\underline{\hphantom{xxxxxx}}$ + $\underline{\hphantom{xxxxxx}}$

5. Model this reaction again, this time using the number of each particle indicated in the chemical equation above. Were you now able to form all the products indicated in the chemical equation?

Summarizing Your Thoughts

6. In some chemical equations a coefficient is not written. What coefficient is implied when a numeral is not written?

7. In a balanced chemical equation, what can be said about the number of atoms of each element in the products and reactants?

8. Describe what it means for a chemical equation to be balanced.

Team Skills

9. As a group, come up with three concepts communicated in today's work to which you will need to dedicate additional study time outside of class in order to become more comfortable with them.

ACTIVITY 10.3 BALANCING CHEMICAL EQUATIONS

Objective

- Be able to use a balanced chemical equation communicate information about the quantities of compounds involved in a reaction

The Model

$$MgCO_3 + 2\,LiF \rightarrow MgF_2 + Li_2CO_3$$

Exploring the Model

1. What does the coefficient of 2 mean in the equation presented as the model?

2. If one unit of magnesium carbonate reacts as shown in the Model, how many lithium fluoride units are needed?

3. How many units of each product will be formed in the reaction if one unit of magnesium carbonate reacts?

4. If ten units of magnesium carbonate react, how many lithium fluoride units will have reacted with them?

5. If ten units of magnesium carbonate react, how many lithium carbonate units will be produced?

6. If one million units of magnesium carbonate react, how many lithium fluoride units will have reacted with them?

7. Explain how you determine the answer to the questions above.

8. If 0.8 moles of magnesium carbonate react, how many moles of lithium fluoride will have reacted with them?

9. How does your approach change when you consider the number of units versus the number of moles?

10. The coefficients present in a balanced chemical equation can be used as conversion factors to calculate how many units of any compound will be formed or used in a chemical reaction. In this case, we can think of 1 unit of $MgCO_3$ = 2 units of LiF (where a unit is any countable number, usually an individual molecule or a mole. Use this relationship to set up a conversion factor that shows how to do the math to solve problem #8:

    0.8 moles $MgCO_3$ x ───────────────── = ___ moles LiF

11. For the following chemical equation, balance the equation, and use the proper ratios of molecules to determine how much aluminum chloride would be formed in this reaction if 100 atoms of aluminum were consumed in the reaction:

    _____ Al + _____ Cl_2 → _____ $AlCl_3$

    100 atoms Al x _____ = _____ molecules $AlCl_3$

Summarizing Your Thoughts

12. In chemistry we use the mole most often to represent the amount of a substance. Why are moles used rather than individual molecules?

13. What information does a balanced chemical equation give that an unbalanced chemical equation does not?

14. In determining how much of a compound is produced in a reaction, why is it important to use a balanced chemical equation?

Team Skills

15. When something was not well understood in this activity, which group member was best able to explain it to the other members of the group? How did they explain the concepts so that the other members of the group understood them better?

16. With this in mind, what can other members of the group do to better understand the material presented, and what study techniques will be most useful in preparing for the examination over this material?

ACTIVITY 10.4 MASS IN CHEMICAL EQUATIONS

Objective

- Be able to use a balanced chemical equation to communicate mass quantities of reactants consumed or products formed in a chemical reaction

The Model

Ted continues to consider his construction of a grilled cheese sandwich, which follows the equation shown below. He recognizes that if he has two slices of bread and one slice of cheese, he can make one grilled cheese sandwich, which is described by the equation.

$$2 \text{ Bd} + \text{Ch} \rightarrow \text{Bd}_2\text{Ch}$$

In thinking about how this relates to chemical reactions, he thinks of a similar chemical equation, as:

$$2 \text{ Na} + \text{S} \rightarrow \text{Na}_2\text{S}$$

Exploring the Model

1. With one slice of cheese, how many sandwiches could be made?

2. With one sulfur atom, how many sodium sulfides could be made?

3. Would it be true to state that with 100 grams of cheese, 100 grams of grilled cheese sandwiches could be produced? Why or why not?

4. What additional information would be needed to answer question #3?

5. Would it be true to state that with 100 grams of sulfur, 100 grams of sodium sulfide could be produced? Why or why not?

6. Since we are unable to work with atoms or molecules individually, we have to deal with them in moles. What value is used to relate the number of moles of sulfur to the mass of sulfur?

7. Calculate the number of moles of sulfur in 100 grams of sulfur, showing your work.

8. With this information, how many moles of sodium sulfide could be produced when 100 grams of sulfur are consumed in the chemical reaction represented?

Summarizing Your Thoughts

9. As a group, discuss and agree on steps that need to be taken to determine how much product is formed when a mass quantity of a reactant is consumed in a chemical reaction. Write the steps below.

End of Chapter Exercises

Activity 10.1

1. Balance each of the chemical equations below, using whole numbers to represent the numbers of each particle that make the numbers of atoms of each element equal on both sides of the equation. (Helpful hints: it is useful to start balancing by starting with the chemical formula with the largest number of atoms, and balancing may start with either a reactant or a product.)

    a. ___ H_2 + ___ O_2 → ___ H_2O

    b. ___ HNO_3 + ___ Cu → ___ $Cu(NO_3)_2$ + ___ H_2

    c. ___ Fe + ___ O_2 → ___ Fe_2O_3

    d. ___ $NaNO_3$ → ___ $NaNO_2$ + ___ O_2

2. The (unbalanced) chemical equation representing water decomposing into its component elements is: H_2O → H_2 + O_2. Write out this chemical equation, and balance it. How is this equation related to the one in 1a above?

3. A (seemingly) easier but incorrect way to do this would be to change the equation to be: H_2O_2 → H_2 + O_2. Why is changing coefficients to balance a chemical equation acceptable, but changing subscripts to balance a chemical equation is unacceptable?

Activity 10.2

4. Below are a series of chemical equations. Place coefficients in front of each chemical formula as needed to balance each chemical equation.

    a. ___ $CaCO_3$ → ___ CaO + ___ CO_2

    b. ___ $NiCl_2$ + ___ $NaOH$ → ___ $Ni(OH)_2$ + ___ $NaCl$

    c. ___ Ag + ___ Cl_2 → ___ $AgCl$

    d. ___ C_2H_2 + ___ O_2 → ___ CO_2 + ___ H_2O

5. Write and balance a chemical equation representing each of the following reactions.

    a.   Calcium hydroxide reacts with hydrogen fluoride to form calcium fluoride and water

    b.   A reaction which occurs when an iron(II) chloride solution and a potassium carbonate solution is mixed yields iron(II) carbonate and a potassium chloride solution.

    c.   Lithium metal and molecular nitrogen combine to form lithium nitride

    d.   Solid iron added to a copper(II) sulfate solution will form solid copper and a solution of iron(II) sulfate

Activity 10.3

6. Balance the chemical equation below representing the combination of aluminum and molecular chlorine to form aluminum chloride. Use this to answer the questions that follow.

    a.   _____ Al + _____ Cl_2 → _____ $AlCl_3$

    b.   How many chlorine molecules would be consumed in this reaction if 100 atoms of aluminum were consumed? Use the ratio from the balanced chemical equation in the space below to show how this number may be calculated.

        100 atoms Al x ————————— = _____ molecules Cl_2

    c.   How many aluminum chloride particles will be produced with the reaction of 100 atoms of aluminum?

7. Magnesium hydroxide is used as an antacid in milk of magnesia and reacts with hydrogen chloride in the stomach to form water and magnesium chloride.

    a.   Write out the chemical equation representing this reaction, and balance the equation.

    b.   If a dose of milk of magnesia contains 0.020 moles of magnesium hydroxide, how many moles of water will be formed when it reacts?

Activity 10.4

8. Balance the chemical equation shown, representing the combustion of methane, the primary component of natural gas. Use this equation to answer the questions which follow, and show your work.

    a. ___ CH_4 + ___ O_2 → ___ CO_2 + ___ H_2O

    b. How many moles of methane (CH_4) are there in 37.2 grams of methane?

    c. How many moles of carbon dioxide would be produced from the combustion of the number of moles of methane calculated in part b?

    d. How many grams of carbon dioxide would be produced from the combustion of the number of moles of methane calculated in part b?

9. Balance the chemical equation shown, representing the formation of ammonia. Use this equation to answer the questions which follow, and show your work.

    a. ___ H_2 + ___ N_2 → ___ NH_3

    b. How many moles of molecular nitrogen are present in a sample of nitrogen with a mass of 264 grams?

    c. How many moles of ammonia would be produced if 264 grams of ammonia reacted with hydrogen as shown?

    d. How many grams of ammonia would be produced if 264 grams of ammonia reacted with hydrogen as shown?

10. Balance the chemical equation shown. Use this equation to answer the questions which follow, and show your work.

   a.   _____ Al $_{(s)}$ + _____ H$_3$PO$_{4\ (aq)}$ → _____ H$_{2\ (g)}$ + _____ AlPO$_{4\ (s)}$

   b. How many moles of molecular hydrogen will be generated with the reaction of 46.3 grams of aluminum?

   c. How many grams of molecular hydrogen will be generated with the reaction of 46.3 grams of aluminum?

   d. How many grams of aluminum phosphate will be generated with the reaction of 46.3 grams of aluminum?

11. Zinc metal reacts with molecular oxygen to form zinc(II) oxide.

   a. Give the properly balanced chemical equation representing this reaction.

   b. How many moles of zinc oxide will be formed from the reaction of 73.1 grams of zinc?

   c. How many grams of zinc oxide will be formed from the reaction of 73.1 grams of zinc?

12. Sulfur trioxide decomposes to form sulfur dioxide and molecular oxygen.

   a. Give the properly balanced chemical equation representing this reaction.

   b. How many moles of molecular oxygen will be formed from the reaction of 27.4 grams of sulfur trioxide?

   c. How many grams of molecular oxygen will be formed from the reaction of 27.4 grams of sulfur trioxide?

11

What's Going On? (Chemical Reactions)

Getting Started

Things are always happening in a chemical reaction. Observations of the world around you in which it looks like nothing is happening may be misleading. Nothing is static in the chemical world. The dynamic nature of chemistry is a big challenge for a textbook. How do we show change when our words and pictures can't change? Well, in this chapter we will ask you to help.

We ask that you find some macroscopic object (a marker) to represent an atom. Your marker item could be small bingo chips, candies (M&Ms™ or Skittles™), or even wadded up pieces of paper. However, what is important as you begin is that you prepare a color-code key so that you can remember which atoms are being represented.

Have fun gaining an understanding the dynamic nature of chemistry!

Chapter Goals

- Understand how chemical equations relate to a particulate-level process

- Be able to define and identify reactants and products in a chemical reaction

- Classify combination and decomposition reactions from chemical equations

- Identify how atoms rearrange in single-replacement reactions

- Explain which subatomic particles are exchanged in single-replacement reactions

- Associate the terms *oxidation* and *reduction* with the transfer of a subatomic particle

- Recognize trends in double-replacement reactions in order to be able to predict the identity of possible products

- Understand how atoms rearrange in combustion reactions

- Link macroscopic to particulate levels of matter

ACTIVITY 11.1 MODELING CHEMICAL REACTIONS

Objectives

- Understand how chemical equations relate to a particulate-level process

- Be able to define and identify reactants and products in a chemical reaction

Getting Started

Chemical reactions are the rearrangement or recombination of the starting material's atoms or molecules (the reactants) into new substances (the products). Observations of chemical reactions can be made on the macroscopic level by visual inspection of a chemical reaction. However, a key aspect to chemistry is the ability to visualize, link, and represent what happens on the particulate level (the individual molecules or atoms) in chemical equations to observations that are made on the macroscopic level.

A symbolic representation of a chemical reaction that uses chemical symbols to represent atoms, elements or compounds and the changes they undergo is used throughout science. But the best chemists will look at the symbolic representation and think about the nanoscale changes that are occurring.

The Model

Below is a *chemical equation*, a symbolic representation of a *chemical reaction*, that uses chemical symbols to represent molecules and the changes they undergo. A particulate-level representation of this reaction is shown in the boxes below:

$$C + O_2 \rightarrow CO_2$$

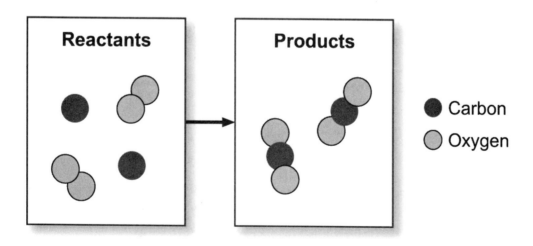

Exploring the Model

1. Give the chemical symbols and names for all the elements represented in the model.

2. Give the chemical formulas and names for all the compounds represented in the model.

3. How are atoms and molecules represented differently in this model?

4. In the model, place a marker representing carbon over the black circles in the box on the left and markers representing oxygen over the other circles in the box on the left. Move the markers to the box on the right, and re-arrange them to form CO_2. After you go through the process outlined above, how many markers are left in the reactant box?

5. Did the number of markers required change as you proceeded? _____

6. How do your observations about the macroscopic changes relate to the changes you observe in the model?

7. As the reaction progresses, what happens to the number of product molecules in the container?

8. As the reaction progresses, what happens to the total number of atoms in the container?

Summarizing Your Thoughts

9. In your own words, what is the definition of a reactant in a chemical reaction?

10. In your own words, what is the definition of a product in a chemical reaction?

11. How is the relationship between reactants and products shown symbolically in a chemical equation?

12. From your observations, explain why the statement, "Matter can't be created or destroyed in a normal chemical reaction," is true.

13. Describe any additional questions about chemical equations that came up in your discussion of this activity. What possible answers to those questions did your group discuss?

Team Skills

14. As a group, imagine a collection of a lot of carbon atoms and oxygen molecules in an airtight container that recombine and become carbon dioxide over the course of several hours. As the reaction progresses (as time passes), what happens to the number of reactant molecules in the container?

15. Write three questions that are related to the chemical reaction that your group still needs to have answered. Discuss these questions as a group, and summarize your discussion.

ACTIVITY 11.2 COMBINATION AND DECOMPOSITION REACTIONS

Objectives:

- Classify combination and decomposition reactions from chemical equations

Getting Started

In Activity 11.1 we looked at a simplified model reaction. For this activity, it will be easiest for you to use the given marker colors to represent each element.

Model

The following is the suggested system used throughout chemistry for coloring models of atoms:

Suggested Marker Color Scheme	Color Scheme Used
Carbon: black	Carbon:
Oxygen: red	Oxygen:
Chlorine: green	Chlorine:
Nitrogen: blue	Nitrogen:
Hydrogen: white	Hydrogen:

Reaction 1: $CO + Cl_2 \rightarrow COCl_2$

Reaction 2: $N_2H_4 \rightarrow N_2 + 2 H_2$

Exploring the Model

1. What is the name of the reactant compound in Reaction 1? _____

2. Sketch and label a nanoscale view of each molecule in Reaction 1 and 2.

3. Draw two boxes on a sheet of loose paper (you will not turn this page in). Label one box "before" and the other "after." Use your markers to create the "before" picture for Reaction 1.

   a. How many atoms are in the "before" box? _____

   b. How many molecules are in the "before" box? _____

4. Rearrange the markers from your Reaction 1 "before" box as you move them to the "after" box.

   a. How many atoms are in the "after" box? _____

   b. How many molecules are in the "after" box? _____

5. Use your markers to create the "before" picture for Reaction 2.

    a. How many atoms are in the "before" box? _____

    b. How many molecules are in the "before" box? _____

6. Rearrange the markers for Reaction 2 as you move them to the "after" box.

    a. How many atoms are in the "after" box? _____

    b. How many molecules are in the "after" box? _____

Summarizing Your Thoughts:

7. Identify which reaction in the model is a combination reaction, and provide a general definition for a combination reaction. Check this definition with your instructor.

8. If a chemical equation were given, what characteristics would you use to determine if the equation represented a combination reaction?

9. Another type of reaction, a decomposition reaction, could be described by the chemical equation $H_2CO_3 \rightarrow H_2O + CO_2$. Which reaction in the model is also a decomposition reaction? In your own words, give the definition of a decomposition reaction.

10. In grammatically correct sentences, write a paragraph comparing and contrasing decomposition reactions and combination reactions.

ACTIVITY 11.3 SINGLE REPLACEMENT REACTIONS

Objectives

- Identify how atoms rearrange in single-replacement reactions

- Explain which subatomic particles are exchanged in single-replacement reactions

- Associate the terms *oxidation* and *reduction* with the transfer of a subatomic particle

Getting Started

Use your markers to help you see that we can replace one atom from a compound, but a new twist has been introduced. You will need to recall information about how to determine the ionic charge on a cation to complete this exercise. Look back to your earlier summaries to see how to determine the charge on a cation.

The Model

Model this reaction with all the necessary reactant particles in the proper grouping in the "reactants" box on the paper, and rearrange the markers to form all the product particles in the "products" box on the paper using the marker colors indicated below for each atom of that element.

Suggested Marker Color Scheme	**Color Scheme Used**
Chlorine: green	Chlorine:
Iron: amber	Iron:
Cobalt: pink	Cobalt:

Reaction 3: $Fe + CoCl_2 \rightarrow Co + FeCl_2$

Exploring the Model

1. How many different colors (representing different elements) are in the "reactants" box before the reaction occurs? _____

2. How many different colors (elements) are in the "products" box after the rearrangement? _____

3. What happens to the chloride ions during this reaction?

4. The iron (Fe) in the Model began as elemental iron with no charge, and possesses a +2 charge in the product. What fundamental particle is associated with this change: the proton, the neutron, or the electron?

5. Did the iron gain or lose the particle(s)? _____

6. How many particles were involved in the change? _____

7. Why are there two chloride ions paired with one cobalt ion in the "reactants" box?

Summarizing Your Thoughts

8. Reaction three represents a single-replacement reaction. Single-replacement reactions commonly occur with ionic compounds and metals. In your own words, what has to happen during the course of a single-replacement reaction?

9. What clues are present in a chemical equation that will help you recognize single-replacement reactions?

10. Oxidation and reduction are opposite processes that must always occur together, involving the exchange of electrons. In the reaction presented in the Model, the cobalt ion is reduced. Based on this information, give a definition of the term reduction.

11. Oxidation occurs when an atom or ion loses electrons through a chemical reaction. In the chemical equation above, which particle is oxidized? _____

12. Why must oxidation and reduction always occur together?

ACTIVITY 11.4 DOUBLE REPLACEMENT REACTIONS

Objective

- Recognize trends in double-replacement reactions in order to be able to predict the identity of possible products

Getting Started

Things are getting a little more complicated, but there are definite trends in this model. You should recall your nomenclature rules and be prepared to recognize some polyatomic ions in order to see these trends.

The Model

Suggested Marker Color Scheme	**Color Scheme Used**
Sodium: purple	Sodium:
Calcium: orange	Calcium:

Reaction 4: $2 \, NaOH + CaCl_2 \rightarrow Ca(OH)_2 + 2 \, NaCl$

Exploring the Model

1. What are the names of the reactant compounds in this reaction?

2. Identify the polyatomic ion in the reaction. _____

3. Model the reactant molecules in the "reactants" box using the markers.

4. Circle the chemical symbols for the cations in the chemical equation. Write the names of the cations present in this reaction.

$$2 \, NaOH + CaCl_2 \rightarrow Ca(OH)_2 + 2 \, NaCl$$

5. Does the charge on each ion change as the reaction proceeds? _____

6. Why is it unreasonable to predict that one of the products would be CaNa?

7. What are the reasonable combinations for the sodium cation given only those ions available as reactants?

8. In a complete sentence, what happens to the polyatomic ion during the course of the reaction? Does it break apart?

Summarizing your Thoughts

9. The Model represents a double-replacement reaction. In your own words, give the definition of a double-replacement reaction.

10. In some reactions, polyatomic ions (such as OH⁻ in the reactions above) can be treated as a group in balancing a chemical equation rather than being treated as separate atoms (O and H). Why might it be beneficial to work with these atoms as a group rather than as individual atoms?

11. Review your exercises, and write a list of clues to look for when trying to predict the types of products that are produced in a double-replacement reaction.

Team Skills

12. There are several possible "jobs" for group members in this activity. Write down the name of each member of the group and what job they did to help the group complete this activity.

ACTIVITY 11.5 COMBUSTION REACTIONS

Objectives

- Understand how atoms rearrange in combustion reactions
- Link macroscopic to particulate levels of matter

Getting Started

This reaction describes the burning of methane (the principal component of natural gas). Methane has a series of compounds call hydrocarbons that behave very similarly. Use your markers to explore this class of chemical reactions.

The Model

A series of combustion reactions:

$$\text{Reaction 5: } CH_4 + 2\,O_2 \rightarrow CO_2 + 2\,H_2O$$

$$\text{Reaction 6: } 2\,C_2H_6 + 7\,O_2 \rightarrow 4\,CO_2 + 6\,H_2O$$

$$\text{Reaction 7: } 2\,C_6H_6 + 15\,O_2 \rightarrow 12\,CO_2 + 6\,H_2O$$

Exploring the Model

1. The hydrocarbons are listed first in each equation. What elements are always present in a hydrocarbon? _____

2. When a hydrocarbon combusts (burns), what does it react with from the air? _____

3. What compounds are produced in all combustion reactions? _____ + _____

4. Why are the coefficients not the same for CO_2 in these reactions?

5. Use your markers to model Reaction 5. Now try modeling this reaction with one oxygen molecule instead of two in the "reactants" box. Can the proper number of product molecules be obtained? _____

Summarizing Your Thoughts

6. Summarize the things you should be able to recognize in order to predict when a reaction is a combustion reaction.

7. If you were given only the formula for a hydrocarbon, describe what other compounds or elements would be needed to predict the correct chemical equation for the combustion of that hydrocarbon.

8. Once you predict the reaction, you need to assure that the number of each element is the same on the reactant and product sides. Describe the steps you will take to assure that this equality is true.

Team Skills

9. One member of your group gets a gold star for their work today. Decide as a group which member deserves the gold star. Why did that person get the gold star today?

10. For each other member of the group, state how that member can improve their participation on these activities to deserve the gold star in a future activity.

End of Chapter Exercises

Activity 11.2

1. What is the difference between a chemical reaction and a chemical equation?

2. Briefly describe why having a standardized color scheme would be helpful to a chemist.

3. Write a short paragraph that explains why coefficients are needed when writing chemical equations.

4. In this reaction,

    $$2\,AgNO_3 + CuCl_2 \rightarrow Cu(NO_3)_2 + 2\,AgCl$$

    a "2" is placed in front of the $AgNO_3$ and the $AgCl$. What does this number represent?

5. Explain why the coefficient "2" was necessary in the chemical equation.

6. Two types of reactions are represented in this activity: combination and decomposition. For each reaction, decide whether or not the reaction is a combination reaction. Then determine whether the equation shown needs to include any coefficients. If coefficients are needed, write them in the appropriate place.

	Combination, decomposition, or neither?	Coefficients needed?
$Hg + S \rightarrow HgS$	_____	_____
$Al + Cl_2 \rightarrow AlCl_3$	_____	_____
$HgO \rightarrow Hg + O_2$	_____	_____
$NH_3 + I_2 \rightarrow NI_3 + H_2$	_____	_____

Activity 11.3

7. Determine which element is oxidized and which is reduced in the following reactions:

   a. $Ni_{(s)} + Cl_{2\ (g)} \rightarrow NiCl_{2\ (s)}$

   b. $3Fe(NO_3)_{2\ (aq)} + 2Al_{(s)} \rightarrow 3Fe_{(s)} + 2Al(NO_3)_{3\ (aq)}$

   c. $Cl_{2\ (aq)} + 2NaI_{(aq)} \rightarrow I_{2\ (aq)} + 2NaCl_{(aq)}$

   d. $PbS_{(s)} + 4H_2O_{2\ (aq)} \rightarrow PbSO_{4\ (s)} + 4H_2O_{(l)}$

Activity 11.4

8. Below are four general equations for combination, decomposition, single-replacement and double-replacement reactions with letters representing different elements. In the blank following each chemical equation, state what type of chemical reaction it best represents.

   $AB + C \rightarrow CB + A$ _____

   $QR \rightarrow Q + R$ _____

   $DE + FG \rightarrow DG + FE$ _____

   $X + Y \rightarrow XY$ _____

9. Write the chemical equation, including coefficients if necessary, for the reaction of calcium hydroxide with iron(II) sulfide to form iron(II) hydroxide and calcium sulfide.

10. Use your markers to model the chemical change in the iron(II) sulfide reaction above; then sketch what the "products" box looks like after the reaction is complete (put the chemical symbol for each atom inside its corresponding circle).

11. In a reaction between calcium chloride and potassium carbonate, what products would be formed?

Activity 11.5

12. Prepare a list of hydrocarbons that are commonly used as fuels. Think about household items that use a fuel. What hydrocarbons are common?

13. Another hydrocarbon that burns is ethene, C_2H_4. Obtain enough markers to produce CO_2 and H_2O with no atoms left over. How many reactant oxygen molecules are needed for this to occur? _____

14. Write a balanced chemical equation for the combustion of ethene.

15. What about the chemical equations describing the combustion of ethene and methane is the same?

16. Compare and contrast the balanced equations for the combustion of ethene and methane.

17. Hydrogen and oxygen gases may react to form water as described by the chemical equation: 2 $H_2 + O_2 \rightarrow H_2O$. Would you consider this a combustion reaction? Why or why not?

18. An environmentalist claims that burning a gallon of gasoline puts five pounds of carbon into the atmosphere. Use the following calculations to evaluate this statement.

      a. Assume that the hydrocarbon octane approximates gasoline. Write the balanced chemical equation for combustion of octane.

      b. Convert one gallon of octane to grams of octane, using an assumed density for gasoline (gasoline floats on water, clearly state your assumed density).

      c. Convert grams of octane to grams of CO_2. Using your knowledge of grams-to-moles calculations and the balanced equation.

      d. Convert grams of octane to grams of C.

      e. Does the statement try to inflate the amount of carbon in the atmosphere by using five pounds of CO_2 and calling it "carbon"?

      f. Make a few assumptions to calculate the average amount of CO_2 that a student generates driving to and from school per year. State all assumptions and show your calculation in a clear, logical manner.

12

How Much Is Possible? (Limiting Reactants & Percent Yield)

Chapter Goals

- Understand what limits the quantity of products

- Be able to define the terms excess reactant and limiting reactant

- Be able to determine the limiting reactant in a chemical reaction, given mole quantities of the reactants

- Be able to determine how much of a compound may be produced in a limiting reactant problem

- State the mass relationships between reactants and products in a chemical reaction

- Be able to determine the limiting reactant in a chemical reaction, given mass quantities of the reactants

- State what percent represents

- Be able to calculate percent yield for a chemical reaction

ACTIVITY 12.1 LIMITING REACTANTS

Objectives:

*   Understand what limits the quantity of products

*   Be able to define the terms excess reactant and limiting reactant

The Model

Jim is the designated grilled cheese sandwich maker at the local sandwich shop, and each day he is given bread and cheese to combine to make grilled cheese sandwiches, represented as:

$$2\ Bd\ +\ Ch\ \rightarrow\ Bd_2Ch$$

On different days, Jim is given different amounts of bread and cheese, and makes the number of grilled cheese sandwiches as shown in Table 12.1.

Table 12.1

Day	Slices of Bread	Slices of Cheese	Sandwiches Made
1	26	10	10
2	18	14	9
3	24	12	12
4	40	17	17
5	32	32	16

Exploring the Model

1.   How many grilled cheese sandwiches could you make from 26 slices of bread?

2.   Why were fewer sandwiches than that made on Day 1?

3.   How many grilled cheese sandwiches could you make with 14 slices of cheese?

4.   Why were fewer sandwiches than that made on Day 2?

Summarizing your Thoughts

5.   The limiting reactant was slices of cheese on Day 1, and slices of bread on Day 2. Using complete sentences and proper grammar, give a definition for a limiting reactant.

6.   The excess reactant was slices of bread on Day 1, and slices of cheese on Day 2. Using complete sentences and proper grammar, give a definition for an excess reactant.

7.   On which day was there no limiting reactant?

Team Skills

8.   Describe how the balanced chemical equation is used as part of your calculation of the maximum amount of product you can make.

ACTIVITY 12.2 MOLE QUANTITIES IN LIMITING REACTANTS

Objectives

- Be able to determine the limiting reactant in a chemical reaction, given mole quantities of the reactants

- Be able to determine how much of a compound may be produced in a limiting reactant problem

The Model

Gaseous oxygen (O_2) and gaseous hydrogen (H_2) combine to form water (H_2O)

Exploring the Model

1. Write out the balanced chemical equation for the reaction in the model.

Prepare a new sheet of paper by drawing two boxes. Label these boxes as "before" and "after." Use bingo chips or candies (Skittles™ or M&Ms™) to represent atoms in the reaction. For the activity, it will be easiest for your group to use the red chips to represent oxygen, and white chips to represent hydrogen. This is the standard system for coloring models of atoms that is used throughout chemistry (but if you can't find white candies make an appropriate substitution).

2. Start with all of the necessary reactants and quantity of each reactant as described by your balanced equation in the before box on the paper. Check your groupings with your facilitator. Rearrange the reactant particles to form all of the product particles in the after box on the paper. Describe whether the quantity of products forms matches what was represented in the chemical equation.

3. Repeat the reaction by modeling this reaction with two oxygen molecules and two hydrogen molecules as reactants. Create as many water molecules as possible. How many of each molecule is present after the reaction is complete?

    Oxygen –

    Hydrogen –

    Water –

4. If two-dozen oxygen molecules were combined with two-dozen hydrogen molecules, what would be the limiting reactant, and how much water could be formed?

5. If two moles of oxygen molecules were combined with two moles of hydrogen molecules, what would be the limiting reactant, and how much water could be formed?

6. Explain how scaling the number of particles from two-dozen to two moles changes the way you think about how many water molecules could be formed.

We rarely actually deal with integer numbers of moles of reactants, which can make determination of a limiting reactant slightly more difficult. To determine which is the limiting reactant, we can use a mathematical approach, in which the quantity of each reactant present can be used to calculate the amount of product formed if all of each reactant is fully consumed. For example, if 7.23 moles of hydrogen and 3.76 moles of oxygen is combined, how much water could be formed?

7. In the space below, fill in the ratio of reactants from the balanced chemical equation in the space below to calculate how much water could be produced if all of the hydrogen present reacts.

    7.23 moles H_2 x ———————————— = _____ moles H_2O

8. In the space below, fill in the ratio of reactants from the balanced chemical equation to calculate how much water could be produced if all of the oxygen present reacts.

    3.76 moles H_2 x ———————————— = _____ moles H_2O

9. From your answers to the questions above, discuss as a group and describe what the limiting reactant is if 7.23 moles of hydrogen and 3.76 moles of oxygen is combined, and water is formed. Explain how you arrived at that answer.

Summarizing your Thoughts

10. Given molar quantities of two reactants in a chemical reaction, how would you determine which is the limiting reactant?

11. How would you determine the quantity of a product that may be formed in a limiting reactant problem?

Team Skills

12. Come to a consensus in your group, and describe how the coefficients in a balanced chemical equation differ from the molar quantities of reactants given in a description of a chemical reaction.

13. Below are four reactions that describe the combustion of four different hydrocarbons.

    $CH_4 + 2\,O_2 \rightarrow CO_2 + 2\,H_2O$

    $2\,C_2H_6 + 7\,O_2 \rightarrow 4\,CO_2 + 6\,H_2O$

    $C_2H_4 + 3\,O_2 \rightarrow 2\,CO_2 + 2\,H_2O$

    $2\,C_2H_2 + 5\,O_2 \rightarrow 4\,CO_2 + 2\,H_2O$

    In a container, 2.6 moles of one of these hydrocarbons combines with 7.8 moles of oxygen and is burned. After the reaction is complete, some hydrocarbon is left in the tank. Which of the above reactions could it be? How did you solve this problem?

Aluminum is gradually added to a container containing a gas, and the amount of product formed is monitored. Figure 12.1 represents the relationship between the aluminum added and the product formed.

Figure 12.1

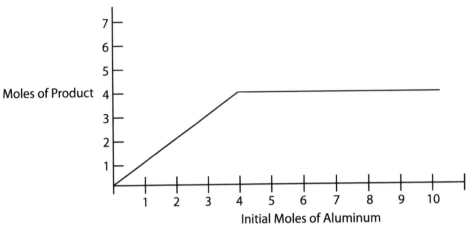

14. As a group, answer the following questions about the reaction shown

   a. After the additon of one mole of aluminum, what is the limiting reactant, and how much product was formed?

   b. After the addition of three moles of aluminum, what is the limiting reactant, and how much product was formed?

   c. After the addition of five moles of aluminum, what is the limiting reactant, and how much product was formed?

   d. After the addition of all of the aluminum (10 moles), how much product was formed?

15. The gas to which the aluminum was added was either nitrogen, oxygen, or fluorine, and reacted in one of the following ways, depending on which gas was present.

$$2\ Al\ +\ N_2\ \rightarrow 2\ AlN$$

$$4\ Al\ +\ 3\ O_2\ \rightarrow 2\ Al_2O_3$$

$$Al\ +\ 3\ F_2\ \rightarrow\ 2\ AlF_3$$

   a. Based on the results shown in Figure 12.1, which gas was present in the container, which reacted with aluminum?

   b. Explain how your group arrived at your answer, and what different results would be expected for the other gases.

16. What was the most difficult part of this activity for your group to complete?

17. What insights did each person in your group have as a result of completing this activity? Give each group member's name and an insight that they had.

Periodic Table of the Elements

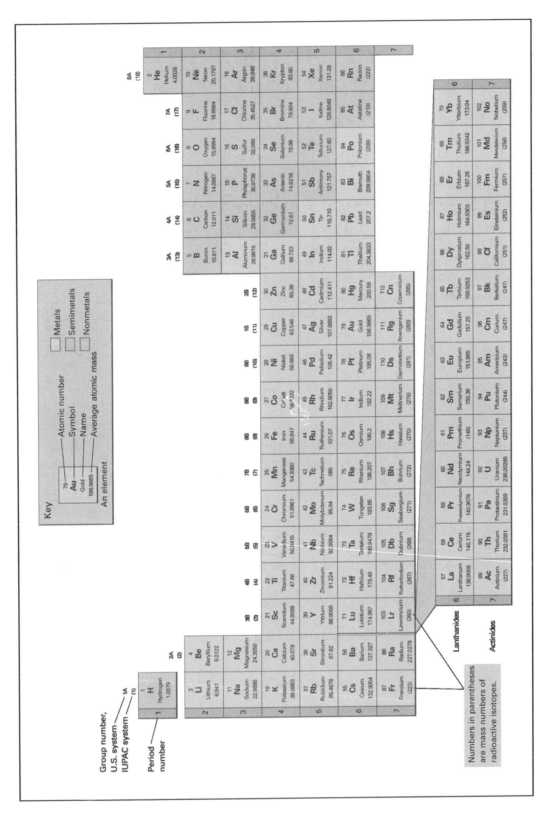

ACTIVITY 12.3 MASS QUANTITIES IN LIMITING REACTANTS

Objectives

* State the mass relationships between reactants and products in a chemical reaction

* Be able to determine the limiting reactant in a chemical reaction, given mass quantities of the reactants

The Model

Debbie carries out a reaction in the lab, in which calcium sulfide is formed from the combination of calcium and sulfur. Debbie wants to produce 50 grams of calcium sulfide, and recognizes the chemical equation describing this reaction as:

$$Ca + S \rightarrow CaS$$

Debbie recognizes since the calcium to sulfur ratio is 1:1 from the equation, she will need 25 grams of each reactant to be consumed in the reaction to make the 50 grams of product. She carries out the reaction, and while 25 grams of calcium is consumed in the reaction, only 45 grams of calcium sulfide has been produced when the reaction is complete.

Exploring the Model

1. Was Debbie correct in her understanding that the calcium and sulfur are consumed in a 1:1 ratio in this reaction?

2. How many moles of calcium are present in 25 grams of calcium?

3. How many moles of calcium sulfide would be produced with the combination of this number of moles of calcium with sulfur?

4. What would be the mass of calcium sulfide produced, with this amount of calcium consumed?

5. How many moles of sulfur will react with this amount of calcium to form calcium sulfide?

Summarizing your Thoughts

6.   What is the limiting reactant in the reaction described in the Model?

7.   Where did Debbie's logic fail in her initial assumptions?

8.   Given mass quantities (in grams) of two reactants in a chemical reaction, how would you determine which is the limiting reactant?

Team Skills

9.   Discuss and come to a group consensus: In a chemical reaction, are each of the following conserved or not between reactants and products: number of molecules in reactants and products, number of atoms of each element, mass of reactants and products?

   a.   number of atoms of each element between reactants consumed and products formed

   b.   number of molecules between reactants consumed and products formed

   c.   mass of reactants consumed and products formed

10. How many grams of sulfur remain after this reaction is complete? How does this relate to the mass of calcium sulfide produced?

ACTIVITY 12.4 PERCENT YIELD

Objectives

- State what percent represents
- Be able to calculate percent yield for a chemical reaction

The Model

Ernie has a solution of made up of 24.3 grams of silver nitrate dissolved in water, and wants to remove the silver from it as a solid. He finds that the silver ions in solution can be turned to solid silver following the equation:

$$2\ AgNO_{3\ (aq)}\ +\ Cu_{(s)}\ \rightarrow\ 2\ Ag_{(s)}\ +\ Cu(NO_3)_{2\ (aq)}$$

Ernie adds copper to his solution, and later filters out silver from his solution container. After drying, he carefully weighs out the silver obtained, and finds that it has a mass of 11.6 grams

Exploring the Model

1. How many moles of silver nitrate are present in Ernie's initial solution?

2. Based on your previous answer, how many moles of silver could be obtained from the solution?

3. How many grams of silver could be obtained from the solution?

4. How does this number compare to the number of grams of silver that Ernie obtained?

1. What percent of the amount of silver that was in the solution was Ernie able to obtain? Check your answer here with your facilitator.

Summarizing your Thoughts

6. The term for the value that you calculated in Question #5 is *percent yield.* Write a definition in your own words for the term *percent yield.*

7. Explain why you can't use grams of product divided by grams of reactant to calculate percent yield.

8. Is it possible to have a percent yield greater than 100%?   Why or why not?

Team Skills

9. Provide *at least* two reasons you think percent yield might be important to a chemist.

10. Which aspect of this activity was most difficult for your group to complete?

11. What did your group do in order to be able to get around this difficulty?

12. List each member of the group. Have each member give two skills that they need to develop (or concepts which they need to work on) in order to master the material presented for the next exam.

End of Chapter Exercises

Activity 12.1

1. At a tricycle factory, a shipment comes in with 400 seats and 600 wheels. Assuming that all other parts are available, how many tricycles can be made?

2. In the tricycle example, what information did you know (but wasn't stated in the problem) which was required to solve the problem?

3. There are more wheels than there are seats, so how can there be seats left over after all of the tricycles are assembled?

Activity 12.2

4. Write out the balanced chemical equation describing a double-replacement reaction between magnesium bromide and lithium hydroxide.

5. If ten moles of each of magnesium bromide and lithium hydroxide are combined and react completely according to your balanced chemical equation, how many moles of lithium bromide will be formed?

6. Write the balanced chemical equation describing a double-replacement reaction between lead(II) nitrate and potassium sulfate.

7. If 1.43 moles of lead(II) nitrate react with an excess amount of potassium sulfate, how much lead(II) sulfate would be formed? The framework is set up below to use the ratios from the balanced chemical equation:

    1.43 moles $Pb(NO_3)_2$ x ——————————————— = _____ moles $PbSO_4$

8. If 0.87 moles of potassium sulfate react with an excess amount of lead(II) nitrate, how many moles of lead(II) sulfate would be formed?

9. If 1.43 moles of lead(II) nitrate and 0.87 moles of potassium sulfate are combined and react as shown, how many moles of sodium carbonate would be formed? What is the limiting reactant in this example?

10. If 3.17 moles of acetylene (C_2H_2) and is combined with 6.82 moles of oxygen in a combustion reaction, how many moles of carbon dioxide would be formed? What is the limiting reactant in this example?

Activity 12.3
11. Mercury is obtained by refining it from the ore cinnabar, which is composed of mercury(II) sulfide. If 329 grams of cinnabar is refined, how many grams of mercury could be obtained?

12. Vinegar contains acetic acid (CH_3COOH) which reacts with baking soda (sodium hydrogen carbonate, $NaHCO_3$), to form water, carbon dioxide gas, and sodium acetate. This reaction can be described by the equation:

$$CH_3COOH_{(aq)} + NaHCO_{3\,(s)} \rightarrow H_2O_{(l)} + CO_{2\,(g)} + NaCH_3CO_{2\,(aq)}$$

   a. If 32.5 grams of baking soda is added to a solution containing 28.9 grams of acetic acid, what is the limiting reactant?

   b. With the reaction of 32.5 grams of baking soda and 28.9 grams of acetic acid, how many moles of carbon dioxide will be produced?

   c. With the reaction of 32.5 grams of baking soda and 28.9 grams of acetic acid, how many grams of carbon dioxide will be produced?

13. Ammonia (NH_3) can be formed from the combination of hydrogen gas and nitrogen gas. If 143 grams of hydrogen gas is placed in a container with 982 grams of nitrogen gas, how many grams of ammonia could be formed?

Activity 12.4

14. A sample of 94.4 grams of lithium carbonate reacts with excess calcium chloride in a double-replacement reaction, and calcium carbonate is collected. How many grams of calcium carbonate would be expected to form in this reaction? (Don't forget to think in terms of moles.)

15. If 123.5 grams of calcium carbonate is actually obtained, what is the percent yield?

16. In a reaction, 25 grams of sodium hydroxide reacts with hydrogen chloride to form water and 25 grams of sodium chloride. Is the percent yield in this process 100%? Describe your thought process.

17. Acetone is a widely used chemical with the chemical formula C_3H_6O. It is commonly made from isopropyl alcohol, C_3H_8O, by removing molecular hydrogen, H_2. A new process for production of acetone is developed, and in order for it to compete with current processes, at least 92% of the expected product must be obtained. If 323 moles of isopropyl alcohol are used in the system, what is the minimum number of moles of acetone that must be obtained for the new process to be competitive?

13

Gases and the Laws

Chapter Goals

- Explain or predict physical phenomena relating to gases in terms of the ideal gas model
- Describe how pressure influences volume (Boyle's law)
- Plot data to visualize relationships between gas variables
- Be able to define inverse and proportional relationships
- Describe how temperature influences volume (Charles's law)
- Describe how the number of gas particles and the volume are related (Avogadro's Law)
- Describe the relationship between the ideal gas law and the other gas laws

ACTIVITY 13.1 INTERACTIONS IN THE GAS PHASE

Objective

- Explain or predict physical phenomena relating to gases in terms of the ideal gas model

Model 1

Imagine that you and a friend have been stranded on a small raft in the middle of the ocean for days.

Figure 13.1 You and a friend in an oval-shaped raft

Exploring the Model

1. Given your current surroundings and situation, do your respective physical sizes (or volumes) have any influence on each other?

2. What is the major factor that allows you to interact with your friend?

Model 2

Imagine that you and your friend get *really* tired of interacting with one another and choose to inflate the extra raft. Your friend climbs into the second raft, and the ocean currents set you drifting in opposite directions.

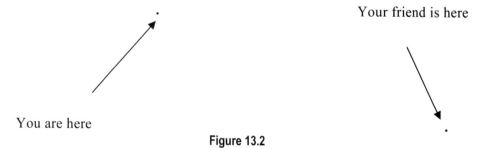

Your friend is here

You are here

Figure 13.2

3. Given your current surroundings and situation, does your physical size (or volume) have any influence on your friend, who is now many, many miles away?

4. If you do not possess any communication devices, what is the major factor preventing you from interacting with your friend or the nearest human?

5. Describe **why** your physical size (or volume) is very significant to you and your friend while in the raft together.

6. Describe **why** your physical size (or volume) when you are alone way out in the middle of the ocean is insignificant to your friend or anyone else.

Summarizing Your Thoughts

7. Recalling that gas particles spread out to fill their container, how would an individual gas particle's physical size (or volume) relate to a sample of gas particles?

ACTIVITY 13.2 BOYLE'S LAW

Objective

- Describe the relationship between volume and pressure (Boyle's law)

Getting Started

As we have seen previously, gases are made of atoms, and therefore have a mass and volume. These properties are important to remember, because we often mistake the small mass of a gas to be "no mass". You should think about how far apart the particles of a gas are (the density) as compared to solids or liquids as you proceed.

For the following examples, the gas particles are depicted as dots. It is important to note that although this may be interpreted as representing individual atoms, most gases occur as molecules, but the same rules apply regardless of the structure of the individual gas particles.

Model 1

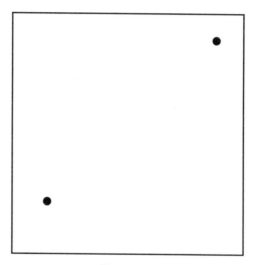

Figure 13.3

Consider the container (box) containing two gas particles (black dots) in Figure 13.3 in answering the following questions.

Model 2

Figure 13.4

The container (box) has now been made much smaller. Note that the gas particle size has not changed.

Exploring the Models

1. In grammatically correct sentence(s), compare the size of the gas particles relative to the size of the container in the two models.

2. Gas particles have considerable kinetic energy, and are in constant motion. In comparing the boxes in the Model, would you expect there to be a difference in the number of times the gas particles collide with each other and the walls of the container? In which figure do you expect to observe more collisions?

3. A simplified way to define pressure is related to the number of collisions per unit area per unit time. Which model would be expected to have a higher pressure?

Summarizing Your Thoughts

4. If the container that holds the gases experiences an external pressure strong enough to force the gas particles so close together that they must interact with one another, what has effectively been removed completely from this sample?

ACTIVITY 13.3 BOYLE'S LAW PART TWO

Objective

- Plot data to visualize relationships between gas variables

- Be able to define inverse and proportional relationships

The Model

Volume and pressure are inversely proportional; $P \propto 1/V$. In other words, as an external pressure increases (becomes larger), the volume occupied by a fixed number of gas particles at constant temperature decreases (becomes smaller). So pressure and volume can contribute in different proportions that always equal a constant value; $P \times V = k$. An initial and final set of circumstances can be equated to one another through this constant value.

$$P_i \times V_i = P_f \times V_f$$

Exploring the Model

1. Use the data below to prepare a graph that shows the relationship indicated in the model. Sketch the pressure versus volume plot in the space provided.

P(mm Hg)	V (mL)
5.0	40
10.0	20
15.0	13.3
17.0	11.8
20.0	10.0
22.0	9.10
30.0	6.70
40.0	5.0

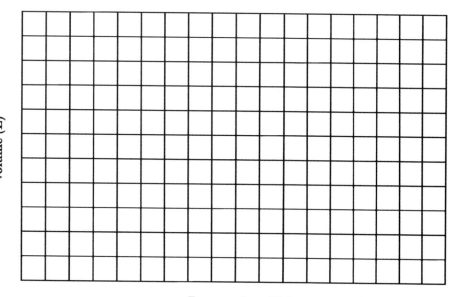

Pressure (mm Hg)

2. Calculate the inverse of volume (1 / Volume) from the values in the table, and plot this new relationship in the space below.

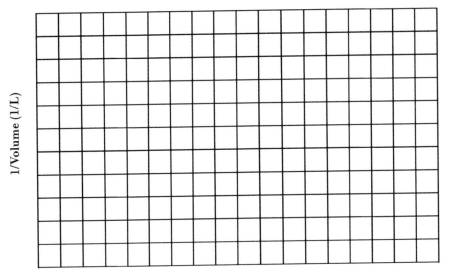

Pressure (mm Hg)

3. Relate the mathematics shown in this activity to the nanoscale view presented in the previous activity.

4. Pressure and volume have an inverse relationship, which is presented in the first plot that you made in this activity. What does the term *inverse relationship* mean?

5. Pressure and the inverse of volume have a direct relationship, which is presented in the second plot that you made in this activity. What does the term *direct relationship* mean?

Summarizing Your Thoughts

6. Explain the usefulness of the linear graph that you prepared. Why would a linear relationship be important for communicating with other people?

7. As a group, discuss other every day, macroscopic observations of gases. Where are some common places that pressurized gases observed?

ACTIVITY 13.4 CHARLES' LAW

Objective

- Describe how temperature influences the volume that a gas may occupy (Charles' law).

Getting Started

Imagine you have just purchased a bouquet of balloons for your New Year's Eve party from the store. You place them in the car and then stop by the grocery store for some cake and ice cream (and maybe a few other items). When you return, your balloons don't look so cheerful. The cool January air had done a trick on them.

In this activity, we will explore why the volume of the balloon changed. Remember to think about how all of those teeny, tiny gas particles (atoms or molecules) as you proceed, and see if you can't return to your party with balloons that resemble those that you purchased.

The Model

The circles in the figure are to be proportional to volume. The dots represent a gas particle. The arrows associated with the gas particles represent the particles velocity. The longer the arrow, the greater the velocity.

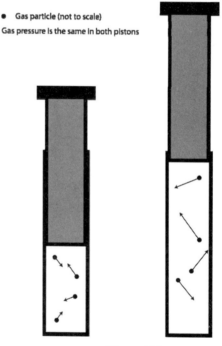

● Gas particle (not to scale)
Gas pressure is the same in both pistons

Figure 13.6

Exploring the Model

1. Use the figure to estimate how the volume changes upon doubling the temperature.

2. In grammatically correct sentences, describe how the velocity changes upon increasing the temperature.

3. Explain why it could be true that the pressure is the same in both views.

Model 2

Use the first trial as a guide to complete the table.

Table 13.2

Trial	Initial Temperature (T_i in Kelvin)	Initial Volume (V_i in Liters)	$V_i/T_i =$	Final Temperature (T_f in Kelvin)	Final Volume (V_f in Liters)	$V_f/T_f =$
1	100	10	10	200	20	10
2	200		0.02	400		0.02
3		8		100		0.02

4. Review the table. Does the ratio of volume to temperature change from the initial condition to the final condition? If the ratio is a constant, could we set the two ratios equal to each other? Using the column headings from Table 13.2, write an equality that is true.

5. Use the relationship to show graphically how temperature and volume are related given a constant mass and pressure.

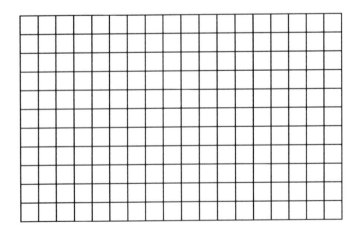

Summarizing Your Thoughts

6. Do temperature and volume exhibit an inverse relationship, or a direct relationship?

7. Write a sentence that describes how temperature and volume are related given constant mass and pressure.

Team Skills

8. As a group discuss how you could build a gas thermometer using the idea that temperature and volume are related. What are the general construction parameters? What are the limitations to the thermometer you could build? Would your thermometer be useful at all temperatures? How big would your thermometer need to be in order to be accurate?

ACTIVITY 13.5 AVOGADRO'S LAW

Objective

• Describe the relationship between the number of gas particles and the volume

The Model

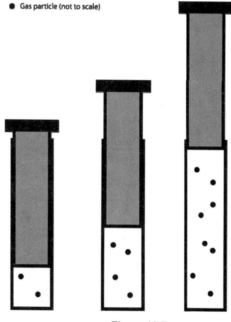

Three identical syringes are shown in Figure 13.7, with different amounts of gas present in each.

Figure 13.7

Exploring the Model

1. Estimate how the volume of the container changes upon the addition of more gas particles at constant temperature and volume, and prepare a plot that represents this relationship.

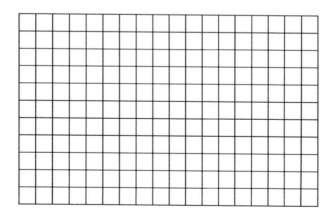

2. Do temperature and volume exhibit an inverse relationship, or a direct relationship?

3. Write this relationship in the form of $y = mx + b$. (What is the expected value of the y-intercept?)

Summarizing Your Thoughts

4. Suppose you observed a gas phase reaction. Explain what you would expect to see if the reaction is a combination reaction. How would the observations be different if the reaction were a decomposition reaction?

5. Explain how knowledge of the relationship between the number of gas particles and the volume would allow you to determine the stoichiometry of a chemical reaction (*i.e.*, explain how you would know what the coefficients would be in the balanced equation).

ACTIVITY 13.6 THE IDEAL GAS LAW

Objective

- Relate the Ideal Gas Law to the other gas laws

The Model

The ideal gas law equation is given as:

$$PV = nRT$$

Where P is pressure, V is volume, n is moles, R is the ideal gas constant, and T is the temperature.

Exploring the Model

1. Show that the ideal gas law is consistent with the statement that pressure and volume have an inverse relationship. What must be held constant for this statement to be true?

2. Rearrange the ideal gas law to isolate pressure on one side of the equation, with all other factors on the other side of the equation. Write your resulting equation below.

3. Rearrange the ideal gas law to isolate temperature on one side of the equation, with all other factors on the other side of the equation. From this equation, state whether temperature has an inverse or a direct relationship with each of the other gas variables:

    Pressure: _____

    Volume: _____

    Moles of gas: _____

4. Consider the ideal gas constant. Describe an experiment that would allow you to determine the value of the ideal gas constant. What needs to be measured? What could easily be held constant?

Summarizing Your Thoughts

5. Describe why learning the Ideal Gas Law makes it unnecessary to memorize the other gas laws. What would you need to be able to do algebraically, and what do you need to know about holding parameters constant in order to apply this knowledge properly?

6. Avogadro's Law states that the volume of a gas increases when the amount of a gas increases. Is it also true that the volume of a gas increases when the amount of a gas increases? Why or why not?

7. Consider a real-life situation in which you use gases. In this situation, which gas variables may change, and which stay the same? Why would it be convenient to work with the Ideal Gas Law which combines all of the gas variables, as opposed to the individual gas laws explored previously?

End of Chapter Exercises

Activity 13.1

1. Conduction is the transfer of heat energy between two substances that are in direct contact. In thinking about the particles that make up matter, why is heat more quickly transferred through as solid than through a gas?

2. Insulated containers like a Thermos® are best able to insulate when there is a vacuum (that is, nothing at all) between the inner container that will hold a hot or cold substance, and the outer container, which is handled. In expanding upon your answer to Question #1, why would a vacuum work better than a gas as an insulator?

Activity 13.2

3. Scuba divers can develop a dangerous condition called "the bends" when they surface too quickly after a deep dive. This occurs because gases dissolved in the blood experience a rapid change from high pressure in deep water to low pressure near the surface. Explain how this is related to Boyle's Law.

4. A closed syringe is an example frequently utilized to show the relationship between pressure and volume. Consider this example to answer the following questions:

    a. If you were able to compress the syringe to a volume of zero, what would that mean that the pressure would need to be?

    b. If you were able to expand the syringe to a pressure of zero, what would the volume need to be?

Activity 13.3

5. Scuba tanks are assigned a size based on the volume of air that they can hold at atmospheric pressure. An 80 cubic foot tank (2265 Liters) therefore can hold an amount of air which would have a volume of 2265 liters at atmospheric pressure. If the air in a tank is pressurized to 204 atmospheres before it is used, what is the actual volume of the tank (in liters)?

6. Given that doubling the pressure reduces a volume of gas to half its original size, fill in the following table:

Table 13.1

Initial Pressure (P_i), atm	Initial Volume (V_i), L	$P_i \times V_i =$ atm ·L	Final Pressure (P_f), atm	Final Volume (V_f), L	$P_f \times V_f =$ atm ·L
1	10		2	5	
	4	4		2	4
2			1		12

Activity 13.4

7. Air bags in a car inflate when a computer in the car detects that it has been in an accident. The amount of nitrogen gas produced in a chemical reaction must be carefully controlled in order for the airbag to inflate properly. What could occur if the amount of nitrogen gas produced is not the proper quantity?

14

Be a Part of the Solution!

Chapter Goals

- Be able to define the terms solution, solute, and solvent

- Be able to differentiate solutions from other classifications of matter

- Know what the term aqueous means

- Determine what ions will be formed in solution when an ionic compound dissolves in water

- Be able to define an dissociation equation

- Be able to state how the charge on an ion plays a part in how it interacts with water molecules in aqueous solution

- Be able to write an ionic equation and net ionic equation describing a chemical reaction

- Be able to define the term spectator ion, and identify spectator ions in a chemical equation

- Be able to give the definition of a precipitate

- Be able to use a particulate-level understanding of solutions to communicate about those solutions

- Be able to combine units to give information about the concentration of a solution

- Know what units of concentration represent

- Be able to use percent-based units of concentration

- Be able to determine when percent-based units of concentration are used

- Know how to use molarity as a unit of concentration

- Be able to determine the number of moles in a given volume of a solution, given the molar concentration of the solution

- Be able to determine the volume of a solution, given the number of moles of solute present in the solution and the molar concentration

- Understand what dilution and concentration mean in dealing with solutions

- Be able to determine the molar concentration of a solution after dilution or concentration

- Be able to give the concentration of component ions in a solution, given the molar concentration of a an ionic compound in the solution

ACTIVITY 14.1 SOLUTION TERMS

Objectives:

- Be able to define the terms solution, solute, and solvent

- Be able to differentiate solutions from other classifications of matter

- Develop a working definition for the term aqueous

The Model

Several mixtures of ethanol and water are made, and data is presented in Table 14.1.

Table 14.1

Amount of Water	Amount of Ethanol	Solvent in Solution	Solute in Solution
173 mL	27 mL	Water	Ethanol
239 mL	319 mL	Ethanol	Water
38 mL	42 mL	Ethanol	Water
94 mL	42 mL	Water	Ethanol
5 mL	17 mL	Ethanol	Water

Exploring the Model

1. For each mixture, what characteristic defines which component is the solvent?

2. For each mixture, what characteristic defines which component is the solute?

3. Which solution contains the largest amount of water? Is water the solvent in this solution?

4. Water and ethanol are *miscible*, which means that they dissolve fully in each other. How does a mixture of water and ethanol differ from a mixture of water and sand?

Summarizing your Thoughts

5. A mixture of water and sand would be heterogeneous, while a mixture of water and ethanol would be homogeneous. For each of the different classifications of matter below, describe what you would expect to see on the particulate level (level of individual atoms or molecules)

      a. heterogeneous mixture -

      b. homogeneous mixture -

      c. compound -

      d. element -

6. Give a definition for the word solvent.

7. Give a definition for the word solute.

8. Give a definition for the word solution.

9. When a substance (or multiple substances) are dissolved in water, it/they may be described as aqueous substances, sometimes communicated with (*aq*) noted after a chemical formula. What is one reason that compounds dissolved in water would have this special designation?

ACTIVITY 14.2 IONIC COMPOUNDS IN AQUEOUS SOLUTION

Objectives

- Determine what ions will be observed in solution when an electrolyte dissolves in water

- Be able to define an dissociation equation

- Be able to state how the charge on an ion plays a part in how it interacts with water molecules in aqueous solution

The Model

$$LiBr_{(aq)} \rightarrow Li^+_{(aq)} + Br^-_{(aq)}$$
$$Ba(OH)_{2\ (aq)} \rightarrow Ba^{2+}_{(aq)} + 2\ OH^-_{(aq)}$$
$$CuSO_{4\ (aq)} \rightarrow Cu^{2+}_{(aq)} + SO_4^{2-}_{(aq)}$$

Exploring the Model

1. What is different about the charges on the reactant and product particles?

2. In each of the above equations, what type of compound (ionic or covalent) is each reactant?

3. What does the (aq) mean after each compound or ion?

4. In each chemical equation, how are the products different from the reactants?

5. Why is there a 2 in front of the hydroxide ions in the second chemical equation?

6. The compounds as shown above dissociate into ions when they dissolve in water, and may be referred to as *electrolytes*. Compounds such as F_2 or CO_2 are *nonelectrolytes*, as they do not dissociate into ions when dissolved in water. How does the type of compound differ between the electrolytes presented here and these nonelectrolytes?

7. Water is a polar compound, meaning that water molecules have a partial positive end, and a partial negative end. The partial positive end of the molecule is in between the hydrogen atoms, with the end of the molecule with the oxygen atom being partially negative. With this molecular polarity, it may dissolve some polar compounds and ionic compounds. The first box below shows how water may surround a bromide ion. In the second box, sketch how water would arrange around a lithium ion in solution.

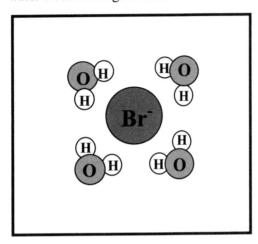 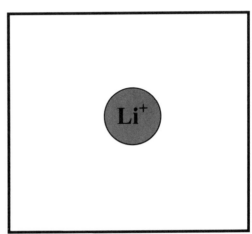

8. How did you decide how to arrange the water molecules around the lithium ion?

9. Complete the following reactions representing the dissociation of ionic compounds in water:

   a. $NH_4Cl_{(aq)}$ → _____ + _____

   b. _____ → $2\ Li^+_{(aq)}$ + $S^{2-}_{(aq)}$

   c. $CaCl_{2\ (aq)}$ → _____ + _____

   d. $Na_2CO_{3\ (aq)}$ → _____ + _____

Summarizing your Thoughts

10. The equations above all represent dissociation equations. In your own words, give a definition for a dissociation equation.

11. In the model, which polyatomic ions are represented?

12. Examine the polyatomic ions, in this activity. Write a rule that states how to include polyatomic ions in dissociation equations.

Team Skills

13. As a team, examine your sketch accompanying Question #7. Use your drawings to explain why ions are frequently soluble in water, while a compound without any charged or partially-charged portions (a nonpolar compound) may not be soluble.

14. In the space below, print the name of each member of your group. This activity contains multiple components based on information that has been covered previously in the course. Have each group member identify a previous topic which was applied in this activity that would be helpful for them to review. Write the identified topic next to each member's name.

Periodic Table of the Elements

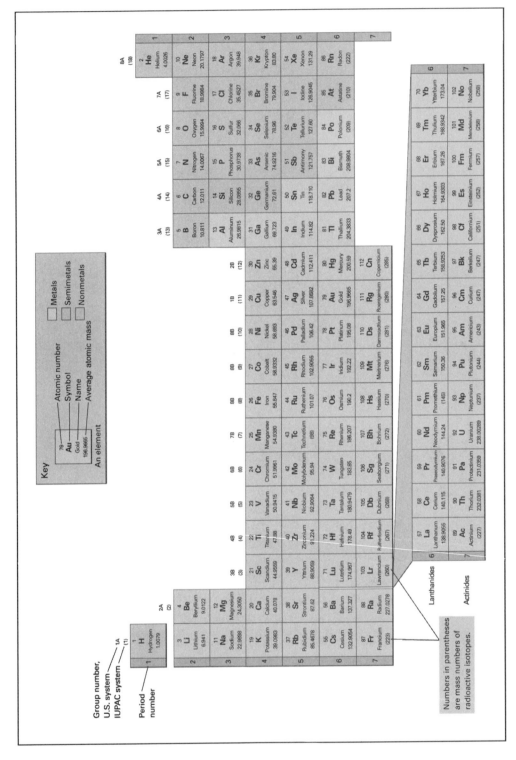

ACTIVITY 14.3 IONIC EQUATIONS AND SOLUBILITY

Objectives

- Be able to write an ionic equation and net ionic equation describing a chemical reaction

- Be able to define the term spectator ion, and identify spectator ions in a chemical equation

- Be able to give the definition of a precipitate

The Model

When aqueous solutions of sodium chloride and silver nitrate are mixed a solid is formed. The observed change can be represented by the following chemical equation:

$$Na^+_{(aq)} + Cl^-_{(aq)} + Ag^+_{(aq)} + NO_3^-_{(aq)} \rightarrow Na^+_{(aq)} + AgCl_{(s)} + NO_3^-_{(aq)}$$

Exploring the Model

1. In the model, how is the identity of the solid indicated?

2. Using the chemical equation presented in the model, write a sentence or two describing what happens to the sodium and nitrate ions upon mixing.

3. At a sporting event, what does a spectator do?

4. In this reaction, the sodium ions and nitrate ions are referred to as *spectator ions*. In a chemical reaction, what does a spectator ion do?

5. This reaction can be summarized with a *net ionic equation*, which would be:

$$Cl^-_{(aq)} + Ag^+_{(aq)} \rightarrow AgCl_{(s)}$$

   How is the net ionic equation different from the equation in the model?

Summarizing your Thoughts

6.   Why would spectator ions frequently not be included in a chemical equation?

7.   In the example in the model, when silver ions and chloride ions combine, they are able to combine to form a solid compound which is no longer soluble in water. As it is formed, it comes out of solution and settles on the bottom of the reaction container.   This is referred to as a *precipitate*. How is this similar to precipitation that you may hear about in weather reports? How are the two definitions different?

8.   What does the behavior of a precipitate tell you about its density, relative to the rest of the solution?

9.   Normally, when a reaction takes place in water (aqueous reactants), water is not included in the chemical equation. When would water need to be included in a chemical equation?

10.   Which of the following terms would be the best description of a solution in which a precipitate has formed, and why? element, compound, homogeneous mixture, or heterogeneous mixture

Team Skills

11.   What concept(s) does your group better understand as a result of today's activity? What concept(s) were explored in today's activity that you have more questions about?

ACTIVITY 14.4 SOLUTION CONCENTRATION

Objectives

- Be able to use a particulate-level understanding of solutions to communicate about those solutions

- Be able to combine units to give information about the concentration of a solution

- Know what units of concentration represent

The Model

A nanoscale view and data of two solutions is shown in Figure 14.1.

Figure 14.1

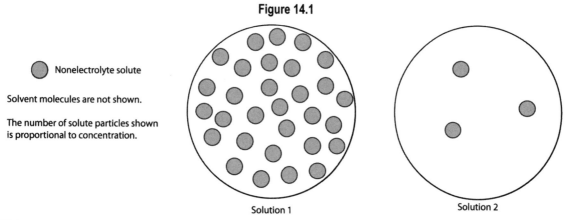

Nonelectrolyte solute

Solvent molecules are not shown.

The number of solute particles shown is proportional to concentration.

Solution 1

Solution 2

	Solution #1	Solution #2
Volume	0.500 Liters	0.500 Liters
Amount of Solute	1 mole	0.1 mole
Concentration	$2 \ ^{moles}/_{Liter}$	$0.2 \ ^{moles}/_{Liter}$

Exploring the Model

1. If the close view circles represent a homogenous mixture in each solution, which solution contains more molecules overall (remember each beaker contains the same volume of liquid)?

2. Which of the solutions would you describe as having a higher concentration of solute molecules?

3. How is the value used for concentration calculated in the solutions above?

4.

Summarizing your Thoughts

4. What could be done to make Solution #2 equal in concentration to Solution #1?

5. A third solution has a volume of 0.750 Liters, and has 0.5 moles of solute dissolved. What is the concentration in moles/Liter of this solution?

6. Units of concentration are expressed as a ratio. Why is it important to express a ratio, and not just how much of a solute is present?

Team Skills

7. Come to a group consensus on the following question: Two solutions, A and B, are held in two separate containers. Is it possible to for Solution A to have less solute, and a higher concentration? Why or why not?

8. In the previous question, who was able to contribute most to the consensus answer for your group? What insight did that individual have which helped your group answer the question?

ACTIVITY 14.5 PERCENT CONCENTRATION UNITS

Objectives

- Be able to use percent-based units of concentration

- Be able to determine when percent-based units of concentration are used

The Model

Table 14.2

Solution	Amount of Solution	Solvent	Solute	Amount of Solute	Concentration of Solute
Rubbing Alcohol	473 mL	isopropanol	water	142 mL	30% by volume
Sterling Silver	500 grams	silver	copper	37.5 grams	7.5% by mass
Saline Solution	355 mL	water	sodium chloride	3.2 grams	0.9% by mass to volume

Exploring the Model

1. What is the solute in sterling silver?

2. What is the solvent in sterling silver?

3. What different units are used to report the amounts of solution and solute in the model?

4. Units expressed as percent deal with a part over the whole. In the model, 142 mL of water is combined with 331 mL of isopropyl alcohol in rubbing alcohol. The calculation of 142 mL / 331 mL does not equal 30% by volume because 331 mL is not "the whole". What is "the whole" in this model? Show the correct calculation.

5. Write a general formula for calculating the percent concentration of a solution for the other two examples.

   a. by mass

   b. by mass to volume

Summarizing your Thoughts

7. In a solution made by mixing two liquids, percent concentration by volume is frequently reported. Why would this unit of concentration be well-suited to mixing two liquids?

8. In a solution made by mixing two solids, percent concentration by mass is frequently reported. Why would this unit of concentration be well-suited to describe a mixture of two solids?

9. For what type of solutions would the percent concentration by mass to volume be most useful?

10. We normally think about solutions as liquid being in the liquid state, such as the case of aqueous solutions. Yet, sterling silver is referred to as a solution. Develop a definition for the term solution that would fit both an aqueous solution and the solid solution (sterling silver).

ACTIVITY 14.6 MOLARITY

Objectives

- Know how to use molarity as a unit of concentration

- Be able to determine the number of moles in a given volume of a solution, given the molar concentration of the solution

- Be able to determine the volume of a solution, given the number of moles of solute present in the solution and the molar concentration

The Model

Beakers containing three different aqueous solutions are shown below.

Figure 14.2

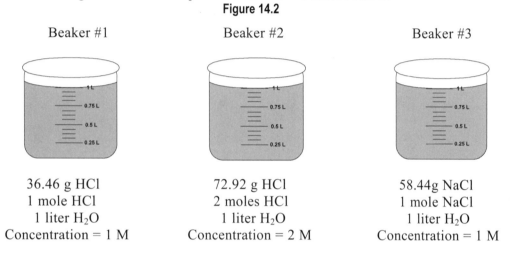

| Beaker #1 | Beaker #2 | Beaker #3 |

36.46 g HCl
1 mole HCl
1 liter H₂O
Concentration = 1 M

72.92 g HCl
2 moles HCl
1 liter H₂O
Concentration = 2 M

58.44g NaCl
1 mole NaCl
1 liter H₂O
Concentration = 1 M

Exploring the Model

1. The molar concentration, or molarity (abbreviated with a capital M) is shown for each of the above solutions. What formula is used to calculate molarity for the solutions above?

2. The solutions in Beaker 1 and Beaker 3 have the same concentration, although Beaker 3 has a greater mass of solute. How is this possible?

3. What units are needed for the amount of a solute in calculating molarity of a solution?

4. What units are needed for the volume of a solution in calculating its molarity?

Summarizing your Thoughts

5. Give the steps necessary to calculate the molar concentration of a solution for which the mass of a solute is given in grams, and the volume of a solution is given in liters.

6. Why would chemists use molarity as a preferred measurement of concentration, instead of other units?

7. If the volume of a solution is given in milliliters, how could the molarity of that solution be calculated?

8. Write a mathematical equation which represents how to determine the number of moles of solute present in a solution, given the solution's molar concentration and volume in liters.

9. Write a mathematical equation which represents how to determine the volume of a solution, if the solution's molar concentration and moles of solute are given.

10. Is it possible for two separate solutions to have a different number of moles of solute and the same concentration? Why or why not?

Team Skills

11. Which question on this activity was most difficult to complete, and which member of the group was most integral in solving this problem? What information did they bring to the table that helped in understanding the problem?

ACTIVITY 14.7 CONCENTRATION AND DILUTION

Objectives

- Understand what dilution and concentration mean in dealing with solutions
- Be able to determine the molar concentration of a solution after dilution or concentration

The Model

Jimmy is thirsty, and makes himself a glass of instant lemonade by following the label instructions. He adds two tablespoons of drink mix concentrate to one cup of water, and stirred until the mix dissolved (solution #1). He takes a drink and decides that it is too sweet, so he dilutes it by adding another ½ cup of water to produce solution #2.

Exploring the Model

1. In going from solution #1 to solution #2, how does the amount of solvent change?

2. In going from solution #1 to solution #2, how does the concentration change?

3. In going from solution #1 to solution #2, how does the amount of solute change?

4. In problems dealing with dilution or concentration of a solution, what value remains the same?

5. Review the everyday use of the terms *dilution* and *concentration*. Which term would most likely be associated with the addition of water shown in the model?

6. The molar concentration of a solution is defined as the number of moles of solute per volume of the solution in liters. Rearrange this equation to put what stays the same in a concentration or dilution problem on one side of the equation, and what changes on the other side.

Summarizing your Thoughts

7. The equation $C_1V_1 = C_2V_2$ is often used in calculations involving concentration or dilution, where the C and V represent concentration and volume before and after the change. In looking at question 6, how can this relationship be derived from the relationship presented there?

8. What does it mean to *dilute* a solution?

9. If you buy orange juice that states it is *from concentrate*, what does that phrase mean?

Team Skills

10. Jimmy is working in the lab later in the day, and measures out 50 mL of 0.30 M NaCl for an experiment, which needs to be transferred to an empty beaker. During the transfer, he spills a small quantity on his lab bench top. If the first circle below represents a particulate-level view of the solution before he spills it, what would the particulate-level view be of the solution spilled on the bench and in the beaker? Come to a group consensus and represent these views in the circles in Figure 14.3.

Figure 14.3

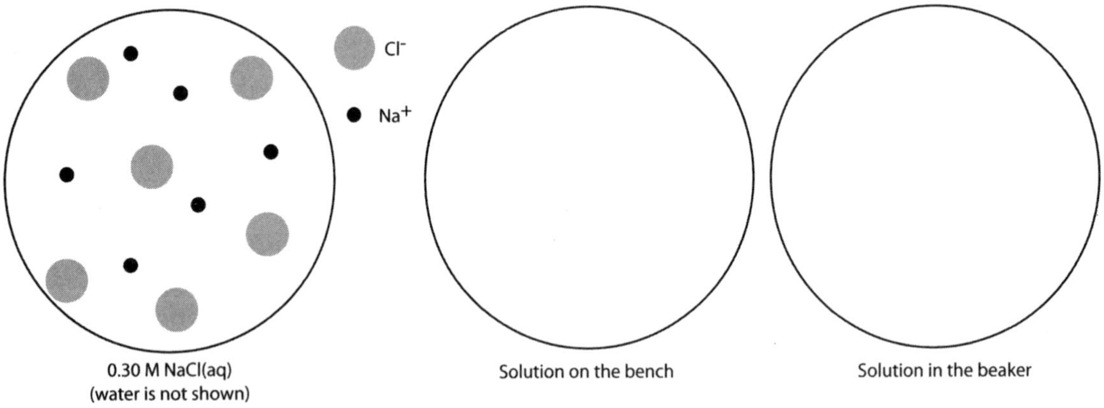

0.30 M NaCl(aq)   Solution on the bench   Solution in the beaker
(water is not shown)

11. Describe in words how the circles above would differ if the beaker to which the solution was transferred was not empty, but instead contained 50 mL of water.

ACTIVITY 14.8 AQUEOUS ION CONCENTRATION

Objectives

- Be able to give the concentration of component ions in a solution, given the molar concentration of a an ionic compound in the solution

The Model

Table 14.3 below gives concentrations of ions present in 1 M solutions of the ionic compounds shown.

Table 14.3

Ionic Compound	Cation	Concentration of Cation	Anion	Concentration of Anion
NaCl	Na^+	1 M	Cl^-	1 M
$CaCl_2$	Ca^{2+}	1 M	Cl^-	2 M
$CuSO_4$	Cu^{2+}	1 M	SO_4^{2-}	1 M
Li_2CO_3	Li^+	2 M	CO_3^{2-}	1 M

Exploring the Model

1. The 1 molar sodium chloride solution is 1 molar in chloride ions, while the 1 molar calcium chloride solution is 2 molar in chloride ions. If the ionic compounds are of the same concentration, how are the ions produced of different concentrations?

2. Give the chemical formula of an ionic compound which would produce 3 moles of chloride ions when one mole is dissolved in water.

3. Given the name of a soluble ionic compound in aqueous solution, how would you determine the concentration of each ion in solution if the concentration of the compound in solution was given? List the steps required.

Summarizing your Thoughts

4.   When oxygen dissolves in water, no ions are formed. Why not?

5.   An dissociation equation describes how an ionic compound dissociates into ions in solution.
      Give the dissociation equation for lithium carbonate dissolving in water. How does this
      representation relate to the concentrations given in the table in this activity?

6.   Two reactions are represented below. What is different about these two reactions?

$$CaCl_2 \rightarrow Ca + Cl_2$$
$$CaCl_2 \rightarrow Ca^{2+} + 2\ Cl^-$$

7.   Which of the previous chemical equations reactions best represents what happens to calcium
      chloride when it is dissolved in water?

8.   What does the other chemical equation represent?

Team Skills

9.   As a group, come up with the three most important concepts which need to be mastered in
      order to be successful in understanding material presented in Chapter 14.

End of Chapter Exercises

Activity 14.1

1. A 59 mL bottle of hand sanitizer contains about 18 mL of water, 37 mL of ethyl alcohol, and 4 mL of fragrances, colorants, and gelling agents. What is the solvent in hand sanitizer?

2. Flavored drink mixes (such as Kool-Aid®) are frequently sold as small pouches, to be mixed to the final drink at home. In making a pitcher of this type of drink, what happens if you do not stir the contents after adding them to water?

3. Which term best describes the unstirred drink mix in water – heterogeneous mixture, homogeneous mixture, compound or element?

4. Which term best describes the drink mix after completely stirring it – heterogeneous mixture, homogeneous mixture, compound or element?

5. List the components of the drink which are in solution after mixing. Identify each as a solvent or solute.

6. The gas mixture that you inhale is approximately 79% N_2, 20% O_2, and 1% other gases. You exhale a mixture of 79% N_2, 16% O_2, 4% CO_2, and 1% other gases. Does the solvent change in this gaseous solution as you breathe in and out? Why or why not?

Activity 14.2

7. State how many of each component ion would be produced from dissociation of each of the following compounds when dissolved in water

    a. potassium iodide

    b. barium nitrate

    c. iron(II) chloride

    d. lithium sulfate

    e. zinc(II) bromide

Activity 14.3

8. Cross out the spectator ions in the following reactions to give the *net ionic equations.*

$$Mg^{2+}_{(aq)} + 2\,Cl^-_{(aq)} + 2\,H^+_{(aq)} + CO_3^{2-}_{(aq)} \rightarrow MgCO_{3\,(s)} + 2\,Cl\text{-}_{(aq)} + 2\,H^+_{(aq)}$$

$$2\,NH_4^+_{(aq)} + S^{2-}_{(aq)} + Fe^{2+}_{(aq)} + 2\,I^-_{(aq)} \rightarrow 2\,NH_4^+_{(aq)} + FeS_{(s)} + 2\,I^-_{(aq)}$$

$$2\,H^+_{(aq)} + SO_4^{2-}_{(aq)} + Ca^{2+}_{(aq)} + 2\,OH^-_{(aq)} \rightarrow 2\,H_2O_{(l)} + CaSO_{4\,(s)}$$

9. Write the *ionic equations* and the *net ionic equations* for the reactions:

    a. Aqueous lead (II) nitrate is combined with aqueous sodium iodide to form solid lead (II) iodide and aqueous sodium nitrate.

    b. Hydrochloric acid is combined with aqueous sodium hydroxide to form aqueous sodium chloride and water.

Activity 14.4

10. In the table, descriptions are given comparing Solution A and Solution B. For each, put an "X" to indicate which solution has the greater concentration for each description, or if this cannot be determined from the information given

Description	Solution A	Solution B	Cannot be Determined
Both solutions have the same volume, Solution A has more solute than Solution B			
Solution A has more solvent and solute than Solution B			
Both solutions have the same amount of solvent, Solution A has more solute than Solution B			
The amount of total solution is the same, Solution A has more solvent than Solution B			
Solution A has less solvent and solute than Solution B			
Solution A has more solvent and less solute than Solution B			

Activity 14.5

11. What is the percent concentration by mass of a solution made by adding 293grams of potassium hydroxide to 2.00 kilograms of water?

12. A 50-pound bag of Portland cement has approximately 2.25 pounds of aluminum oxide. What is the percent composition by mass of aluminum oxide is in Portland cement?

13. A hummingbird feeder solution has a sucrose concentration of 20% by weight to volume. If you would like to make 2 liters of hummingbird food, how much sucrose would be required?

Activity 14.6

14. What is the molarity of a solution made by adding 0.34 moles of magnesium chloride to water to produce 2.31 liters of solution?

15. What is the molar concentration of a solution made by adding 100 grams of lithium hydroxide to water to produce 0.600 L of solution?

16. How many moles of ammonia are present in 0.850 Liters of a 0.5 M ammonia solution?

17. How many grams of calcium fluoride are present in 3.45 Liters of a 0.2 M calcium fluoride solution?

18. What is the total volume of solution required to prepare a 0.2 M solution of potassium bromide when 137 grams of the salt is placed in a beaker?

Activity 14.7

19. Water is allowed to evaporate from 340 mL of a 0.32 M solution of calcium bromide. How many moles of calcium bromide would be left in the beaker?

20. If water is added to the container mentioned above (after the water has evaporated) to make 500 mL of solution what would be the molarity of this new solution?

21. A 0.31 M aqueous solution of ammonium hydroxide solution with a volume of 528 mL is diluted with the addition of 832 mL of water. What is the concentration of the resulting solution?

22. A 2.37 M solution of magnesium sulfate has a volume of 382 mL. A 1.50 M $MgSO_4$ solution is needed for an experiment, which can be obtained by diluting the existing solution. If water is added to the exitsting solution to form a new solution with the desired concentration, what will the new volume of the solution be?

Activity 14.8

23. Complete the table below, giving the expected concentrations of each anion and cation in a 1 molar solution of the given ionic compound.

Ionic Compound	Cation	Concentration of Cation	Anion	Concentration of Anion
Li_2SO_4				
nickel (II) fluoride				
potassium nitrate				
lithium iodide				
cobalt (II) sulfate				
barium hydroxide				

Periodic Table of the Elements

Group number,
U.S. system
IUPAC system

Period number

Key

79
Au
Gold
196.9665

Atomic number
Symbol
Name
Average atomic mass

An element

Metals
Semimetals
Nonmetals

Numbers in parentheses are mass numbers of radioactive isotopes.

Lanthanides

Actinides

1A (1)	2A (2)	3B (3)	4B (4)	5B (5)	6B (6)	7B (7)	8B (8)	8B (9)	8B (10)	1B (11)	2B (12)	3A (13)	4A (14)	5A (15)	6A (16)	7A (17)	8A (18)
1 H Hydrogen 1.0079																	2 He Helium 4.0026
3 Li Lithium 6.941	4 Be Beryllium 9.0122											5 B Boron 10.811	6 C Carbon 12.011	7 N Nitrogen 14.0067	8 O Oxygen 15.9994	9 F Fluorine 18.9984	10 Ne Neon 20.1797
11 Na Sodium 22.9898	12 Mg Magnesium 24.3050											13 Al Aluminum 26.9815	14 Si Silicon 28.0855	15 P Phosphorus 30.9738	16 S Sulfur 32.066	17 Cl Chlorine 35.4527	18 Ar Argon 39.948
19 K Potassium 39.0983	20 Ca Calcium 40.078	21 Sc Scandium 44.9559	22 Ti Titanium 47.88	23 V Vanadium 50.9415	24 Cr Chromium 51.9961	25 Mn Manganese 54.9380	26 Fe Iron 55.847	27 Co Cobalt 58.9332	28 Ni Nickel 58.693	29 Cu Copper 63.546	30 Zn Zinc 65.39	31 Ga Gallium 69.723	32 Ge Germanium 72.61	33 As Arsenic 74.9216	34 Se Selenium 78.96	35 Br Bromine 79.904	36 Kr Krypton 83.80
37 Rb Rubidium 85.4678	38 Sr Strontium 87.62	39 Y Yttrium 88.9059	40 Zr Zirconium 91.224	41 Nb Niobium 92.9064	42 Mo Molybdenum 95.94	43 Tc Technetium (98)	44 Ru Ruthenium 101.07	45 Rh Rhodium 102.9055	46 Pd Palladium 106.42	47 Ag Silver 107.8682	48 Cd Cadmium 112.411	49 In Indium 114.82	50 Sn Tin 118.710	51 Sb Antimony 121.757	52 Te Tellurium 127.60	53 I Iodine 126.9045	54 Xe Xenon 131.29
55 Cs Cesium 132.9054	56 Ba Barium 137.327	71 Lu Lutetium 174.967	72 Hf Hafnium 178.49	73 Ta Tantalum 180.9479	74 W Tungsten 183.85	75 Re Rhenium 186.207	76 Os Osmium 190.2	77 Ir Iridium 192.22	78 Pt Platinum 195.08	79 Au Gold 196.9665	80 Hg Mercury 200.59	81 Tl Thallium 204.3833	82 Pb Lead 207.2	83 Bi Bismuth 208.9804	84 Po Polonium (209)	85 At Astatine (210)	86 Rn Radon (222)
87 Fr Francium (223)	88 Ra Radium 227.0278	103 Lr Lawrencium (260)	104 Rf Rutherfordium (267)	105 Db Dubnium (268)	106 Sg Seaborgium (271)	107 Bh Bohrium (272)	108 Hs Hassium (270)	109 Mt Meitnerium (276)	110 Ds Darmstadtium (281)	111 Rg Roentgenium (280)	112 Cn Copernicium (285)						

Lanthanides (6)

57 La Lanthanum 138.9055	58 Ce Cerium 140.115	59 Pr Praseodymium 140.9076	60 Nd Neodymium 144.24	61 Pm Promethium (145)	62 Sm Samarium 150.36	63 Eu Europium 151.965	64 Gd Gadolinium 157.25	65 Tb Terbium 158.9253	66 Dy Dysprosium 162.50	67 Ho Holmium 164.9303	68 Er Erbium 167.26	69 Tm Thulium 168.9342	70 Yb Ytterbium 173.04

Actinides (7)

89 Ac Actinium (227)	90 Th Thorium 232.0381	91 Pa Protactinium 231.0359	92 U Uranium 238.0289	93 Np Neptunium (237)	94 Pu Plutonium (244)	95 Am Americium (243)	96 Cm Curium (247)	97 Bk Berkelium (247)	98 Cf Californium (251)	99 Es Einsteinium (252)	100 Fm Fermium (257)	101 Md Mendelevium (258)	102 No Nobelium (259)

15

Acids and Bases

Chapter Goals:

- Be able to identify neutralization reactions and the products of a neutralization reaction

- Be able to identify acids and bases in a neutralization reaction

- Be able to give the names aqueous acids, given the chemical formulas or names of ionic compounds which may act as acids

- Be able to describe how strong and weak acids differ in their behavior in aqueous solution

- Be able to write an ionization equation describing the behavior of an acid in aqueous solution

- State what the pH and pOH of a solution represent

- Be able to apply characteristics of acids to the pH of an aqueous solution of that compound

- Be able to calculate the pH or pOH of a solution, given the hydronium ion concentration or hydroxide ion concentration of the solution

ACTIVITY 15.1 NEUTRALIZATION REACTIONS

Objective

- Be able to identify neutralization reactions and the products of a neutralization reaction
- Be able to identify acids and bases in a neutralization reaction

The Model

$$HBr + NaOH \rightarrow H_2O + NaBr$$
$$2\,HNO_3 + Ca(OH)_2 \rightarrow H_2O + Ca(NO_3)_2$$
$$HClO_4 + KOH \rightarrow H_2O + KClO_4$$

Exploring the Model

1. What product is common to all of the reactions in the model?

2. Examine the chemical equations written in the model. The first reactant listed in each chemical equation have something in common, what is it?

3. The first reactant in each chemical equation is an acid. As the reaction goes from reactants to products, what happens to the ions that make up an acid?

4. The second reactant in each chemical equation is a base. What do all of the bases have in common?

5. The reactions above are called *neutralization reactions*. What are the terms for the reactants in a neutralization reaction?

6. In a neutralization reaction, the products are always water and a salt. Write the chemical formula for the salt for each of the reactions shown in the Model.

Summarizing Your Thoughts

7. Given the chemical formula of a compound, how could you identify if it was an acid or a base?

8. A neutralization reaction is a specific type of double-replacement reaction, in which reactants switch cations and anions. Describe how cations and anions are exchanged to yield the products in a neutralization reaction.

9. Write out a "generic" form of a neutralization reaction, in which atoms that are always present in a neutralization reaction are stated, and particles which are not always the same are represented by another letter, such as "X" or "Y".

10. In a neutralization reaction, it may be beneficial to think of water as hydrogen hydroxide. Why does this description of water work well for a neutralization reaction?

ACTIVITY 15.2 NAMING AQUEOUS ACIDS

Objective

- Be able to give the names aqueous acids, given the chemical formulas or names of ionic compounds which may act as acids

The Model

Acid solutions (the ionic compounds which act as acids, when dissolved in water) have different names than the ionic compounds themselves. A table of acidic compounds and the names of aqueous solutions which refer to dissolving each in water is shown below.

Table 15.1   Acidic Compounds and Aqueous Acids

Chemical Formula	Ionic Compound Name	Aqueous Acid Name
HF	hydrogen fluoride	hydrofluoric acid
HCl	hydrogen chloride	hydrochloric acid
HBr	hydrogen bromide	hydrobromic acid
HNO_2	hydrogen nitrite	nitrous acid
HNO_3	hydrogen nitrate	nitric acid
H_2SO_3	hydrogen sulfide	sulfurous acid
H_2SO_4	hydrogen sulfate	sulfuric acid
H_2CO_3	hydrogen carbonate	carbonic acid

Exploring the Model

1. What do the ionic compound names of all of the acids have in common?

2. What do the aqueous acid names of all of the acids have in common?

3. What three word endings are seen in the ionic compound names?

    a.

    b.

    c.

4. What do these different word endings tell you about the compounds they represent?

    a.

    b.

    c.

5. What are the aqueous acid name endings which come from each of the ionic compound name endings?

    a.

    b.

    c.

6. In what types of compounds does the aqueous acid name start with the hydro- prefix?

7. How are the number of hydrogen ions determined for each chemical formula?

8. Do the number of hydrogen ions present give different ionic compound names?

Summarizing Your Thoughts

9. Identify the set of rules for how to name an aqueous acid, starting from the chemical formula of a compound. List the rules in the space below.

10. The acid present in vinegar is acetic acid. How would you determine the chemical formula of acetic acid from the name, and what additional information would you need to do this?

Team Skills

11. Identify three things that your team might do to work more effectively and efficiently.

ACTIVITY 15.3 STRENGTH OF ACIDS

Objectives

- Be able to describe how strong and weak acids differ in their behavior in aqueous solution

- Be able to write an ionization equation describing the behavior of an acid in aqueous solution

The Model

Reactions of two acids with water are shown below

$$\text{(1)} \quad HNO_{3\,(aq)} + H_2O_{(l)} \rightarrow H_3O^+_{(aq)} + NO_3^-_{(aq)}$$
$$\text{(2)} \quad HNO_{2\,(aq)} + H_2O_{(l)} \rightleftharpoons H_3O^+_{(aq)} + NO_2^-_{(aq)}$$

Exploring the Model

1. What is different about the first reactant in equations 1 and 2?

2. What is different about the arrows used in equations 1 and 2?

3. There are several ways that acids may be grouped, one of which is strong acids and weak acids. Strong acids, when added to water, fully dissociate into its component anion, with the cation (the proton, H^+) adding to a water molecule to form the hydronium ion (H_3O^+). The top chemical equation represents how a strong acid acts when added to water. What is the name of this aqueous acid?

4. Draw a nanoscale representation of the dissolution of nitric acid in water on the basis of the description above.

5. In weak acids, the other grouping, usually only a small percentage of the acid molecules dissociate into ions, and most of the particles stay as molecules in solution. This is also a reversible process, if the dissociated anion comes in contact with a hydronium ion, the acid molecule can re-form. What is the name of the aqueous acid in the second chemical equation?

6. How is this reversibility of the process communicated in chemical equation 2?

Summarizing Your Thoughts

7. Use the chemical equations presented in the model and the definition given for a strong and weak acid to describe the differences at the nanoscale that are expected when one mole of HNO_3 is added to one liter of water, and one mole of HNO_2 is added to a liter of water in another container. Which solution will have more *ions* in solution?

8. How did you decide on the answer above?

9. Of these two solutions, which will have more acid *molecules* in solution?

10. How did you arrive at this answer?

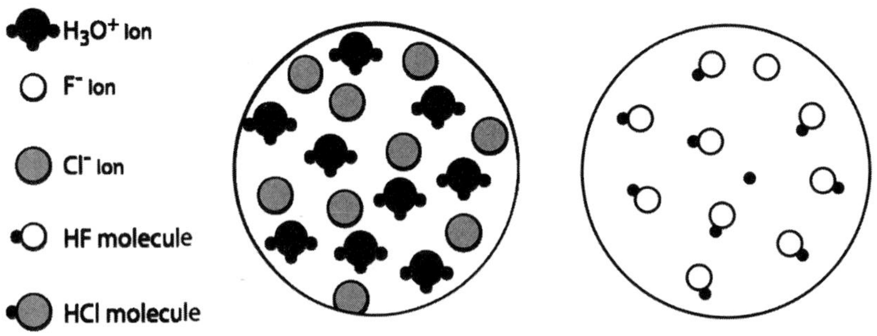

Figure 15.1   **Nanoscale view of HCl and HF**

- H_3O^+ Ion
- F^- Ion
- Cl^- Ion
- HF molecule
- HCl molecule

solutions   water molecules are not shown in this drawing

11. Two acid solutions are depicted above, as what we would imagine that these solutions would look like on the nanoscale. The water is not shown, but only the solute molecules and ions. The image on the left represents hydrochloric acid, and the image on the right represents hydrofluoric. Is hydrochloric acid a strong or weak acid?

12. From the image, how were you able to make this statement about hydrochloric acid?

13. Is hydrofluoric acid a strong or weak acid?

14. From the image, how were you able to make this statement about hydrofluoric acid?

ACTIVITY 15.4 THE pH SCALE

Objectives

- State what the pH and pOH of a solution represent

- Be able to apply characteristics of acids to the pH of an aqueous solution of that compound

- Be able to calculate the pH or pOH of a solution, given the hydronium ion concentration or hydroxide ion concentration of the solution

The Model

Table 15.2 shows the pH of several solutions, made by adding the compound listed to water, to produce a solution of the concentration shown.

Table 15.2   Concentrations of HCl and NaOH solutions, and the resulting pH of the solution

Compound in Solution	Concentration	pH
HCl	0.1 molar	1
HCl	0.01 molar	2
HCl	0.00001 molar	5
-- no compound added, pure water --		7
NaOH	0.000001 molar	8
NaOH	0.001 molar	11
NaOH	0.1 molar	13

Exploring the Model

1. What is the unit used to represent concentration in the model, and what is the formula for that unit?

2. The pH of a solution can generally be stated as how acidic or basic a solution is. As a solution becomes more acidic (the concentration of acid increases), what happens to the pH?

3. As a solution becomes more basic, what happens to the pH?

4. HCl is a strong acid. As HCl is added to water, what two aqueous ions are formed?

5. NaOH is a strong base. As NaOH is added to water, what two aqueous ions are formed?

6. Express the concentration of HCl in solution in the first three rows of Table 15.2 in scientific notation. How are these numbers in scientific notation related to the reported pH of each of these solutions?

Summarizing Your Thoughts

7. pH is defined as an algebraic function of the hydronium ion concentration, mathematically represented as $pH = -\log[H_3O^+]$, where $[H_3O^+]$ represents the molar concentration of the hydronium ion. Why does the pH value change when the amount of acid in a solution changes?

8. The value pOH is used to describe how basic a solution is, based on the concentration of hydroxide ions in a solution. This is very similar to pH describing how acidic a solution is, based on the concentration of hydronium ions. What would be the mathematical representation of pOH?

9. pH also may be described algebraically as $pH = 14 - (-\log[OH])$. Write a mathematical expression showing how pH and pOH are related. Why does the pH value change when the amount of base in a solution changes?

10. A pH value of 7 for a solution is considered to be neutral, or $[H_3O^+] = [OH^-]$. Write out a balanced chemical equation showing how hydronium ions react with hydroxide ions.

11. When an equal number of hydroxide ions react with hydronium ions, a neutral solution is formed. Why is this solution not considered to be acidic or basic?

12. A 0.1 molar hydrochloric acid solution has a pH of 1.00, while a 0.1 molar hydrofluoric acid solution has a pH of 2.10. If these two acids have the same concentration, why is there a difference in the hydronium ion concentration of these solutions?

Team Skills

13. If a hydrochloric acid solution is diluted, how would you expect the pH to change: increase, decrease, or stay the same, and why?

14. If a sodium hydroxide solution is diluted, how would you expect the pH to change: increase, decrease, or stay the same, and why?

15. Look over a group evaluation question from Activity 15.2. Have you changed the way your group works to improve, and address these observations? If so, what has been changed, and if not, is it worth changing in the way suggested, why or why not?

End of Chapter Exercises

Activity 15.1

1. In the list below, circle all of the bases

    Mg(OH)$_2$ H$_2$CO$_3$ CsOH   NaOH   HI

2. In the list below, circle all of the acids

    HCN   MgCO$_3$ HCl   LiOH   H$_2$SO$_4$

3. In the list below, circle all all of the sets of compounds which could react in a neutralization reaction

    HI and Ba(OH)$_2$

    HCl and HNO$_3$

    H$_3$PO$_4$ and KOH

    H$_2$O and KBr

    HF and LiOH

Activity 15.2

4. Complete the table below, with the proper chemical formulas, ionic compound names, and/or aqueous acid names.

Chemical Formula	Ionic Compound Name	Aqueous Acid Name
HClO$_2$	hydrogen chlorite	
H$_2$CrO$_4$		chromic acid
	hydrogen hydroxide	hydrohydroxic acid
HI		
H$_3$AsO$_3$		arsenous acid
H$_2$C$_2$O$_4$	hydrogen oxalate	

Activity 15.3

5. Hydrogen perchlorate ($HClO_4$) is a strong acid. In the space below, write a chemical equation describing how hydrogen perchlorate will behave when added to water.

6. Hydrogen chlorate ($HClO_3$) is a weak acid. In the space below, write a chemical equation describing how hydrogen chlorate will behave when added to water.

7. When HBr is dissolved in water, no HBr molecules exist in solution. Based on this statement, is hydrogen bromide a weak acid or a strong acid, and why?

Activity 15.4

8. Complete the following table with the missing values

$[H_3O^+]$	$[OH^-]$	pH	pOH	Acidic, Basic, or Neutral
		6.0		
0.0063				
	1.0×10^{-7}			
			10.4	
	0.000055			
2.7×10^{-4}				
		9.3		
			4.7	